The Coliform Index and Waterborne Disease

The Coliform Index and Waterborne Disease

Problems of microbial drinking water assessment

Cara Gleeson and Nick Gray

Trinity College
University of Dublin

CRC Press
Taylor & Francis Group
Boca Raton London New York

CRC Press is an imprint of the
Taylor & Francis Group, an **Informa** business

A CHAPMAN & HALL BOOK

CRC Press
Taylor & Francis Group
6000 Broken Sound Parkway NW, Suite 300
Boca Raton, FL 33487-2742

First issued in paperback 2019

© 1997 by Taylor & Francis Group, LLC
CRC Press is an imprint of Taylor & Francis Group, an Informa business

Typeset in 10/12pt Palatino by Acorn Bookwork, Salisbury, Wilts

No claim to original U.S. Government works

ISBN-13: 978-0-419-21870-8 (hbk)
ISBN-13: 978-0-367-86630-3 (pbk)

A catalogue record for this book is available from the British Library
Library of Congress Catalog Card Number: 96–070578

**Visit the Taylor & Francis Web site at
http://www.taylorandfrancis.com**

**and the CRC Press Web site at
http://www.crcpress.com**

Contents

Preface

For nearly a century indicator organisms, in particular the coliform group, have been used to ensure the microbial quality of drinking water. These bacteria were used to indicate the presence of faecal contamination of drinking water and hence the possibility of the presence of waterborne pathogens. *Escherichia coli* had been quickly identified as the preferred indicator, but at that time it could not be specifically enumerated and so an alternative test was developed. We now know this as the total coliform test which not only isolates *Escherichia coli*, which is always found in human and animal faeces, but many other genera such as *Klebsiella*, *Enterobacter*, *Citrobacter* and *Serratia* which are widely found in the environment and are not associated with faecal contamination and do not normally pose a threat to human health. Many of these heterotrophic bacteria also colonize drinking water distribution systems and can rapidly grow resulting in elevated total coliform counts in water, even though no bacterial pathogens are present and so there is no threat to health. Soon after the introduction of the total coliform test a second test was adopted to confirm the presence of faecal coliforms, that is, coliforms associated specifically with faecal contamination. Incubation at 44.5 °C inhibits the growth of non-thermotolerant coliforms except, it was thought, *E. coli* which is thermotolerant. However, other bacteria including *Klebsiella*, a very common bacterial genus, are also thermotolerant. The test is also subject to numerous interferences and problems, with many strains of *E. coli* themselves either unable to ferment lactose or non-thermotolerant resulting in false-negative results. Currently worldwide legislation to protect consumers from microbially unsafe drinking water is based on these outdated and unreliable tests, and while there is considerable concern amongst scientists about their continued use, the water industry and the regulators continue to place almost total reliance on the coliform index. This has serious implications for public health, as increased pressures on diminishing water supplies necessitates the ability to accurately determine water quality.

In the past decade there has been a rapid increase in waterborne outbreaks associated with viral and protozoan agents, normally in drinking waters that were found to be microbially safe using the coliform index. It is now clear that a negative total and faecal coliform result can no longer be taken as ensuring pathogen free water.

In this text we examine the use of the coliform group as indicator organisms. Given recent technological advances in detection techniques, and also what appears to be the inability of the coliform index to adequately indicate the more significant causative agents of waterborne disease in the developed world, the text explores the future use of coliforms.

Chapter 1 establishes the public health significance of waterborne disease, examines the sources of waterborne disease and the risk of infection, and outlines the legislation currently in place with regard to water quality and coliform counts. Chapter 2 defines the concept of indicator organisms and discusses the origin and use of coliforms as indicators. The standard methods used for coliform enumeration are described in Chapter 3, together with their advantages and disadvantages. These methods have long being recognized as having serious limitations and have prompted research in two directions. The first is the development of alternative techniques for the isolation and enumeration of coliforms and *E. coli* which are discussed in Chapter 4. The second is the establishment of alternative indicator systems, the possibilities of which are outlined in Chapter 5. Of increasing concern for water supplies is the emergence in recent years of so-called new pathogens, which are more resistant to water treatment processes than conventional pathogens. These are described in Chapter 6 along with the opportunistic bacterial pathogens such as *Campylobacter* spp. and *E. coli* 0157:H7, as well as the primary pathogens associated with drinking water. Finally, Chapter 7 evaluates the role of the 'coliform count' in future water quality analysis and suggests how existing practice can be improved.

We hope that you find this short text thought provoking.

Cara Gleeson and Nick Gray

Trinity College, Dublin
January 1996

Acknowledgements

We are extremely grateful to all our colleagues who have made material so freely available to us during the preparation of this text. We would also like to thank the following copyright holders for permission to reproduce in full, and to reproduce in amended form, various figures and tables:

Academic Press
  Figures 2.3, 2.4;
American Society of Microbiology
  Figures 5.2, 6.4, 6.5;
  Tables 2.2, 2.3, 3.8, 3.9, 4.1, 5.3;
American Public Health Association
  Figure 6.3;
  Tables 2.1, 3.1, 3.5, 4.2, 6.4;
American Water Works Association
  Figures 5.1, 6.1;
  Table 1.6;
Blackwell Scientific Publications Ltd
  Tables 2.6, 4.3, 4.4, 6.6;
CRC Press Inc.
  Table 5.7;
Elsevier Science Ltd
  Figures 2.2, 6.6;
  Tables 1.1, 1.8, 1.9, 2.5, 5.4, 5.5, 5.8;
  List of factors favouring Cryptosporidiosis in section 6.2;
Epidemiologic Reviews
  Table 1.2;
European Union
  Table 1.15;
The Controller of Her Majesty's Stationery Office
  Figure 6.2;
  Tables 1.10, 1.13;
Institution of Water and Environmental Management
  Figure 1.4;
  Tables 1.2, 1.4, 1.11, 1.12;

National Environmental Health Association
   Figure 2.1;
   Table 2.4;
WCB Publishers Inc.
   Figures 3.1, 4.1, 4.2;
US Environmental Protection Agency
   Tables 1.16, 6.5, 6.7;
John Wiley and Sons Inc.
   Figures 1.1, 1.2;
   Tables 1.3, 5.1, 5.6, 6.2, 6.3;
World Health Organization
   Tables 1.4, 1.5, 1.7, 1.14.

List of abbreviations

A1 medium	broth used for the cultivation of coliforms by the MPN method
APHA	American Public Health Association
BGLBB	Brilliant Green Lactose Bile Broth
cfu	colony forming units
EC broth	*Escherichia coli* broth
ELISA	Enzyme Linked Immuno-Sorbent Assay
EMB agar	Eosin Methylene Blue agar
EU	European Union
GUD	β-D-glucuronidase
HPC	Heterotrophic Plate Count
IBDG	Indoxyl-β-D-Glucuronide
IEA	Immuno-Enzyme Assay
IF	Immuno-Fluorescent Techniques
LES-ENDO	(ENDO agar, LES) ENDO agar, Lawrence Experimental Station
LTB	Lauryl Tryptose Broth
LTLB	Lauryl Tryptose Lactose Broth
MAC	Maximum Admissible Concentration
MCL	Maximum Containment Level
MCLG	Maximum Containment Level Goal
m-CP	*Clostridium perfringens* agar (modified version)
m-ENDO agar	ENDO agar (modified version)
m-ENDO-LES agar	ENDO-LES agar (modified version)
MF	Membrane Filtration
m-FC agar	Faecal Coliform agar (modified version)
m-LSB agar	Lauryl Sulphate Broth (modified version)
MMGM	Minerals Modified Glutamate Medium
MoAb	Monoclonal Antibody
MPN	Most Probable Number
MUG	4-methylumbelliferyl-β-D-glucuronide
ONPG	o-nitro-phenyl-β-D-galactopyranoside
P–A technique	Presence–Absence technique
PCR	Polymerase Chain Reaction

PE Broth	Preston Enrichment Broth
pfu	plaque forming unit
PNPG	p-nitrophenyl-β-D-galactopyranosidase
RIA	Radio-Immuno Assay
SLSB	Sodium Lauryl Sulphate Broth
US EPA	United States Environmental Protection Agency
WHO	World Health Organization
X-gluc	5-4-chloro-3-indoxyl-β-D-glucuronide

Microbial water quality 1

1.1 INTRODUCTION

Water is essential to support life and therefore every effort should be made to achieve a drinking water quality as high as practicable. Failure to do so exposes the population to the risk of disease, particularly the very young, the elderly, the sick and those who live in sub-standard sanitary conditions (WHO, 1993). Because of the potential consequences of waterborne diseases, microbial contamination is still considered to be the most critical risk factor in drinking water quality (Fawell and Miller, 1992). This fact is borne out by reported evidence of waterborne derived illness and disease where the number of reported outbreaks attributable to chemical contamination of drinking water supplies is negligible when compared to the number due to microbial agents (Galbraith, Barnett and Stanwell-Smith, 1987; Herwaldt et al., 1992). Diseases due to microbial agents include acute gastro-enteritis, giardiasis, ameobiasis, cryptosporidiosis, shigellosis and hepatitis, as well as typhoid fever and salmonellosis.

Current strategies for controlling health risks posed by microbes in drinking waters are based on a barrier approach involving treatment of wastewaters as well as the treatment of raw waters which includes disinfection (Figure 1.1); and secondly the establishment of allowable limits for indicators of water quality (Sobsey et al., 1993). From the earliest days of water bacteriology it was recognized that monitoring for the presence of specific pathogens in water supplies was difficult and largely impractical (Bonde, 1977). For this reason a more indirect approach is adopted where water is examined for indicator bacteria whose presence in water implies some degree of contamination. Several organisms have been suggested as potential indicator organisms, but over the years the coliform group has become universally adopted.

The use of indicator organisms, in particular the coliform group, as a means of controlling the possible presence of pathogens has been

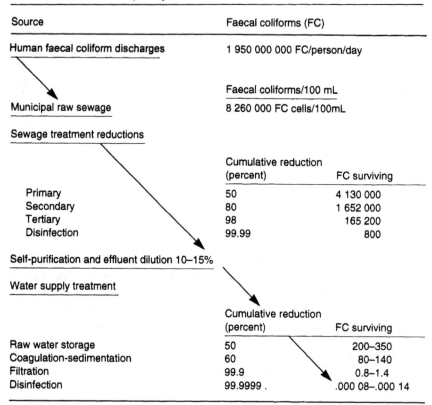

Source	Faecal coliforms (FC)	
Human faecal coliform discharges	1 950 000 000 FC/person/day	
	Faecal coliforms/100 mL	
Municipal raw sewage	8 260 000 FC cells/100mL	

Sewage treatment reductions

	Cumulative reduction (percent)	FC surviving
Primary	50	4 130 000
Secondary	80	1 652 000
Tertiary	98	165 200
Disinfection	99.99	800

Self-purification and effluent dilution 10–15%

Water supply treatment

	Cumulative reduction (percent)	FC surviving
Raw water storage	50	200–350
Coagulation-sedimentation	60	80–140
Filtration	99.9	0.8–1.4
Disinfection	99.9999 .	.000 08–.000 14

Figure 1.1 The use of barriers is vital in the control of pathogens in water supplies (Geldreich, 1991).

paramount in the approach to assessing water quality adopted by the World Health Organization (WHO), US Environmental Protection Agency (US EPA) and the European Union (EU) (EC, 1980; US EPA, 1992; WHO, 1993). This approach is based on the assumption that there is a quantifiable relationship between indicator density and the potential health risks involved. A water quality guideline is established (Figure 1.2) which is a suggested upper limit for the density of an indicator organism above which there is an unacceptable risk (Cabelli, 1978).

Ideally water destined for human consumption should be free from micro-organisms; however, in practice this is an unattainable goal. The concept of allowable risks will be discussed later in the context of risk assessment (section 1.6).

The extent to which strategies, such as the barrier approach and the establishment of allowable limits for bacteria in water, have been successful in maintaining water quality is seen by the dramatic decline

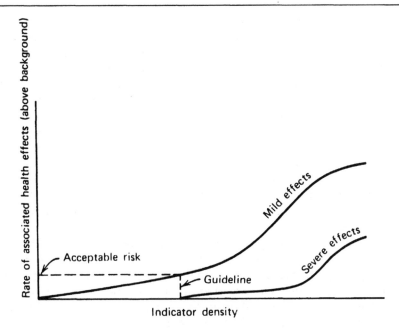

Figure 1.2 The desired water quality criteria and the derivation of guidelines from them (Cabelli, 1978).

in epidemic and endemic waterborne bacterial diseases such as typhoid fever and cholera in the more developed regions of the world (Sobsey *et al.*, 1993). However, despite such success there is increasing dissatisfaction with the use of coliforms as indicator organisms. Since its introduction in the late 1880s, the coliform index has remained a major parameter in water quality standards. Little has changed in this time except for the development and the subsequent incorporation into standards of the membrane filter method and a number of culture media modifications.

Waterborne diseases are now known to be caused by a broader spectrum of microbes than just enteric bacteria. Recent years have seen increasing reports of waterborne outbreaks largely as a result of protozoan and viral agents in waters considered safe to drink under current legislation, which relies largely on the coliform index. This has serious implications for public health as increased pressures on diminishing water supplies necessitates the ability to accurately determine water quality. In addition, there is now higher consumer expectation for quality water supplies. In the UK there has been considerable concern that consumers appear to have lost faith in the safety of water supplies, a sentiment which has been expressed across Europe. Despite the fact that the use of coliforms has been questioned by many workers through the technological and theoretical advances made in public health microbiology, their use has continued and these advances have not been applied to practical use. Dutka (1973) observed that 'hesitancy often results in the adoption of procedures for the sake of security and such procedures may be difficult to get rid of when they have lost all importance'. Has this become the case with the coliform concept?

In addition to the problems encountered with the use of coliforms as indicator organisms, the standard methods currently recommended for their detection are not without their criticisms. The multiple tube method has been used since the turn of the century, while membrane filtration is a relatively more recent development. There is growing concern that neither method accurately reflects the number of coliforms present in a given sample and therefore it is difficult to have full confidence in the results they give.

Such limitations have prompted research in two directions. The first is the development of alternative methods for coliform enumeration. This is perhaps in recognition of the reluctance to change from using coliforms because of their long tradition of use and the extensive data base that now exists in relation to them. The second area of development is the establishment of an alternative indicator system. Since the concept of indicator systems was first developed many organisms have been suggested as potential alternatives to the coliform group and while many of these efforts have proved fruitless others show considerable potential, particularly where the coliform is of recognized limited value such as in tropical climates.

1.2   THE EVOLUTION OF SANITARY SCIENCE

Records and remains indicate an appreciation of the need to protect the quality of water intended for human consumption in even the most ancient of civilizations. Down through the centuries, increased urbanization has placed greater importance on the provision of public water supplies which were of obvious benefit in distributing water to large numbers of people, but these waters were often unsuitable for human consumption. In the mid 1800s, Edwin Chadwick published his now famous report on *The Sanitary Condition of the Labouring Population of Great Britain*. The central theme of this report was the association of disease and death with the insanitary conditions that prevailed in the poorer quarters of cities and towns, and the conviction that 'what was required' was 'drainage, proper cleansing, better ventilation and other means of diminishing atmospheric impurities' (Moore, 1971). Initially, little notice was taken of Chadwick's conclusions and, inevitably with growing urban populations, watercourses and aquifers became increasingly contaminated with sewage (Tebutt, 1992). It wasn't until 1854 that cholera was convincingly linked with the consumption of contaminated waters finally proving that public water supplies could be a source of infection for humans. By removing the handle of the Broad Street pump in London, John Snow brought to an end an epidemic which had claimed almost 500 lives. The well supplying the pump was in close proximity to a sewer (Dadswell, 1990b). This association between disease and contaminated waters focused attention on the public health aspects of water quality and provided the essential evidence for Sedgewick (USA), Chadwick (UK) and Petinkoffer (Germany) who were campaigning for the provision of safe drinking water as a means of disease prevention. It was subsequently found that with better water supplies and sewerage systems, there was a sharp fall in the incidence of waterborne disease in Britain, Europe and North America, even before the agents of waterborne disease had been identified in the laboratory (Olson and Nagy, 1984).

The first microbiological study of drinking water was conducted on a London water supply in 1850 by Hassel, who later published a report entitled *A Microscopic Examination of the Water Supplied to the Inhabitants of London and the Greater Districts*. In this report, Hassel established a link between sanitary quality and microbiological quality. Similar conclusions were also reached by Cohn in 1853 and by other scientists in the latter half of the nineteenth century. During this same period, the germ theory of disease became firmly established as a result of research by Louis Pasteur, Robert Koch and others. The idea of specific microbial indicators was developed in the 1880s by Von Fritsch and later by Escherich (Olson and Nagy, 1984).

By 1900 the concept of waterborne disease was well accepted, the microbial aetiology for many of the diseases which followed an anal–oral route of transmission had been established and the health hazards associated with human consumption had been proven beyond reasonable doubt. Considerable emphasis was placed on recognition of faecal pollution. Initially, early workers depended on sanitary surveys to evaluate faecal contamination of water supplies. These surveys included information on sources of water, topography and geology of the area, soil characteristics and possible sources of contamination. Later, as analytical methodology developed, faecal pollution was evaluated by laboratory methods (Pipes, 1982b). At the close of the nineteenth century there was great diversity in the methods used by sanitary bacteriologists. The American Public Health Association appointed a committee that submitted a report of recommended procedures in 1897. This subsequently led to the publication in 1901 of the first edition of *Standard Methods* (Wolf, 1972).

While considerable emphasis was placed upon recognizing faecal pollution, the importance of water purification techniques was also acknowledged. This was in recognition of the fact that not all source waters were free from contamination and therefore could become a potential public health hazard (Olson and Nagy, 1984). Early treatment facilities such as slow sand filtration produced a significant decline in waterborne disease incidents. A study of cholera by Koch in 1892 provides some of the best evidence of the importance of filtration. During the 1892 cholera epidemic, he traced the incidence of cholera through the cities of Hamburg and Altona. Both cities received their drinking water from the River Elbe, but the City of Altona used sand filtration as the water extracted from the river was known to be contaminated. The results of this study showed that Altona, even with an inferior water source, had a markedly lower incidence of cholera than Hamburg (Safe Drinking Water Committee, 1977). The introduction of chlorination in 1908 further increased the safety of water supplies and the multiple barrier concept of disease prevention was born (McDonald and Kay, 1988). Since the turn of the century there have been many refinements in water purification methods but they are still based on original basic concepts.

1.3 WATERBORNE OUTBREAKS

The reporting of waterborne disease is generally voluntary. In the USA, information is largely obtained from scientific and medical reports, and through the assistance of local and state health officials. Since 1971, the United States Environmental Protection Agency (US EPA) and the Center for Disease Control have co-operated in the investigation and reporting of waterborne outbreaks, and publish outbreak data annually. Statistical data on waterborne outbreaks have been available in the USA since 1920 (Figure 1.3) (Craun, 1991; Singh and McFeters, 1992). The data shows a significant decrease in waterborne disease outbreaks towards the middle of the century. The period 1970 to 1980 saw a rapid increase in outbreaks back to the former level recorded in the 1940s. However, a significant decline has been reported in the five years up to 1990. A breakdown of waterborne outbreaks in the USA between 1920 and 1988 is given in Table 1.1, where the number of outbreaks and cases for each disease is summarized.

Figure 1.4 shows a similar situation for the UK. The initial fall in notifications can be attributed to the effectiveness of water purification methods such as the introduction of chlorination, with subsequent improvements in water supplies, hygiene and sanitation (Galbraith, Barnett and Stanwell-Smith, 1987). Dramatic decreases in typhoid fever and other waterborne diseases accompanied the adoption of chlorination of drinking water supplies (McFeters and Camper, 1983). It is estimated that, from 1946 to 1960, only 1.4% of total typhoid morbidity

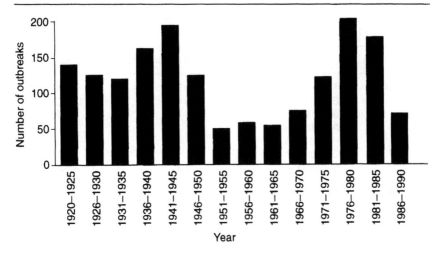

Figure 1.3 Number of waterborne outbreaks per five-year cycle in the USA from 1920 to 1992 (derived from Craun, 1988 and Herwaldt *et al.,* 1992).

could be ascribed to waterborne infection compared to an estimated 40% in 1908 (Moore, 1971).

The increasing number of waterborne disease outbreaks reported in the 1970s and 1980s can be attributed to more aggressive investigation and reporting rather than to a real increase in the number of cases. In recent years, reported waterborne disease outbreaks have tended to occur in small community and non-community water systems, which affect fewer people but mean a larger number of outbreaks are reported each year (Craun, 1977; 1988). Other factors which may account for the

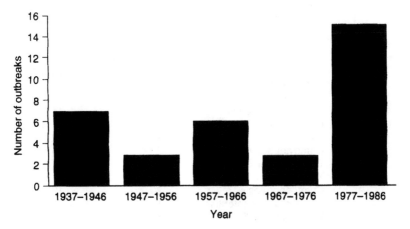

Figure 1.4 Number of waterborne outbreaks in the UK 1937–87 (adapted from Galbraith, Barnett and Stanwell-Smith, 1987).

Table 1.1 Number of waterborne disease outbreaks and cases in the USA between 1920 and 1988 (adapted from Craun, 1991)

Year periods	Disease	Number of outbreaks	Number of cases
1920–40	Typhoid fever	372	13 767
	Gastroenteritis	144	176 725
	Shigellosis	10	3 308
	Amoebiasis	2	1 416
	Hepatitis A	1	28
	Chemical poisoning	1	92
	Subtotal	530	195 336
1941–60	Gastroenteritis	265	54 439
	Typhoid fever	94	1 945
	Shigellosis	25	8 951
	Hepatitis A	23	930
	Salmonellosis	4	31
	Chemical poisoning	4	44
	Paratyphoid fever	3	19
	Amoebiasis	2	36
	Tularaemia	2	6
	Leptospirosis	1	9
	Poliomyelitis	1	16
	Subtotal	424	66 426
1961–70	Gastroenteritis	39	26 546
	Hepatitis A	30	903
	Shigellosis	19	1 666
	Typhoid fever	14	104
	Salmonellosis	9	16 706
	Chemical poisoning	9	46
	Toxigenic *E. coli*	4	188
	Giardiasis	3	176
	Amoebiasis	3	39
	Subtotal	130	46 374
1971–88	Gastroenteritis	279	64 965
	Giardiasis	103	25 834
	Chemical poisoning	55	3 877
	Shigellosis	40	8 806
	Viral gastroenteritis	26	11 799
	Hepatitis A	23	737
	Salmonellosis	12	2 370
	Campylobacteriosis	12	5 233
	Typhoid fever	5	282
	Yersiniosis	2	103
	Cryptospordiosis	2	13 117
	Chronic gastroenteritis	1	72
	Toxigenic *E coli*	1	1000
	Cholera	1	17
	Amoebiasis	1	4
	Subtotal	563	138 216

increase include increased pressures on diminishing water supplies (Pipes, 1982a,b), the use of remote and untreated waters for recreation (Olson and Nagy, 1984) and increased international travel (Galbraith, Barnett and Stanwell-Smith, 1987). In the 1980s, over 85% of all typhoid cases in Britain were contracted abroad (Galbraith, Barnett and Stanwell-Smith, 1987). There has also been the emergence of new bacteria, viruses and protozoans, many of which are resistant to conventional chlorination and have only relatively recently been implicated in waterborne outbreaks (Craun, 1988; Herwaldt *et al.*, 1992).

It should be noted that data on waterborne disease outbreaks should be interpreted with caution as often they do not reflect the true incidence of waterborne disease. Many factors can influence the degree to which outbreaks are recognized and reported in a given year, including interest in the problem and the degree of interest expressed at both local and state level. Craun (1992) estimates that less than 50% and possibly up to 90% of all cases go unreported. This is particularly the case for outbreaks occurring in non-community water systems and individual systems (Herwaldt *et al.*, 1992).

In the UK most outbreaks of waterborne disease are identified by general practitioners, although consumer complaints may also alert water companies of potential outbreaks. The companies themselves may identify that a potential microbial hazard has arisen through their routine surveillance. Alternatively, Environmental Health Officers may identify an outbreak by following up general complaints. Another important source of information of the occurrence of waterborne diseases is the diagnostic microbiology laboratories which, like the general practitioners, report outbreaks directly to their relevant local Public Health Officer known generally as the Consultant in Communicable Disease Control (CCDC) in England and Wales, or to the Director of Public Health. In Scotland the relevant officer is the Consultant in Public Health Medicine. In England and Wales reports are recorded and monitored by the Public Health Laboratory Service (PHLS) Communicable Disease Surveillance Centre (CDSC) and in Scotland by the Communicable Disease and Environmental Health (Scotland) Unit (CDEHU). The CCDC or equivalent gives medical advice relating to waterborne outbreaks and advises as to what further investigations are required. The first action is always to inform the water company involved and if an outbreak is confirmed then the company must inform the Drinking Water Inspectorate who are responsible for investigating all breaches of water quality regulations in England and Wales. The District Health Authority notifies the Department of Health. The CCDC has the executive responsibility during the outbreak and coordinates control measures in close association with the local public health laboratory. Depending on the severity of the outbreak the

Regional Epidemiologist and even the CDSC or CDEHU may be asked for assistance. Guidelines have been issued by the UK Government to ensure that clear lines of command and communication exist in the case of a waterborne outbreak and to ensure direct liaison between all those involved in its control (Department of the Environment, 1994a). Specific action taken to control waterborne outbreaks is discussed in detail elsewhere (WAA, 1985; 1988).

1.4   PUBLIC HEALTH SIGNIFICANCE

Much of the world's population remains without access to potable water supplies and adequate sanitation (Table 1.2) (Esrey and Habicut, 1986). The World Health Organization estimates that 80% of all sickness in the world is attributable to inadequate water supplies or sanitation (Morrison, 1983). Some 250 million new cases of waterborne diarrhoea are reported worldwide each year resulting in more than 10 million deaths (Snyder and Merson, 1982).

There are now many recognized waterborne pathogens. They are all present in large numbers in human or animal excreta, occasionally both, and are generally resistant to environmental decay. Many of these pathogens are capable of causing infections even when ingested in very small numbers.

There are three different groups of micro-organisms that can be transmitted via drinking water: viruses, bacteria and protozoa. They are all transmitted by the faecal–oral route and so largely arise either directly or indirectly by contamination of water resources by sewage or possibly animal wastes. It is theoretically possible, but unlikely, that other pathogenic organisms such as nematodes (roundworm and hookworm) and cestodes (tapeworm) may also be transmitted via drinking water. Table 1.3 lists the common bacterial, viral and protozoan diseases generally transmitted via drinking water.

Table 1.2 Percentage of population covered by adequate water supplies and sanitation facilities (Esrey and Habicut, 1986)

Region	Water supplies		Sanitation	
	Rural	Urban	Rural	Urban
Africa	29	57	18	58
Asia and the Pacific	44	67	9	48
China	40	50		
Latin America and the Caribbean	49	85	20	80
Western Asia	80	95	25	93
Global totals	41	71	12	89

Table 1.3 Bacterial, viral and protozoan diseases generally transmitted by contaminated drinking water (adapted from Singh and McFeters, 1992)

Agent	Disease	Incubation time
BACTERIA		
Shigella spp.	Shigellosis	1–7 days
Salmonella spp.		
S. typhimurium	Salmonellosis	6–72 hours
S. typhi	Typhoid fever	1–3 days
Enterotoxigenic		
Escherichia coli	Diarrhoea	12–72 hours
Campylobacter spp.	Gastro-enteritis	1–7 days
Vibrio cholerae	Gastro-enteritis	1–3 days
VIRUSES		
Hepatitis A	Hepatitis	15–45 days
Norwalk-like agent	Gastro-enteritis	1–7 days
Virus-like particles <27 nm	Gastro-enteritis	1–7 days
Rotavirus	Gastro-enteritis	1–2 days
PROTOZOA		
Giardia lamblia	Giardiasis	7–10 days
Entamoeba histolytica	Ameobiasis	2–4 weeks
Cryptosporidium	Cryptosporidiosis	5–10 days

Infections related to water may be classified into four main groups.

1. **Waterborne disease:** this is where a pathogen is transmitted by ingestion of contaminated water. Cholera and typhoid fever are the classical examples of waterborne disease and this text will, for the most part, be restricted to this form of disease.
2. **Water-washed disease:** these include faecal–orally spread disease or disease spread from one person to another facilitated by a lack of an adequate supply of water for washing. Many diarrhoeal diseases, as well as diseases of the eyes and skin, are transmitted in this way.
3. **Water-based infections:** these diseases are caused by pathogenic organisms which spend part of their life cycle in aquatic organisms. These include schistosomes and other trematode parasites which parasitize snails and guinea worms. An example is dracunculoasis which is spread through minute aquatic crustaceans.
4. **Water-related disease:** caused by insect vectors which breed in water, such as mosquitoes which spread malaria and filariasis arthropods which carry viruses such as those causing dengue and yellow fever (Bradley, 1993).

Table 1.4 Aetiology of waterborne disease in the UK compared over the decades 1937–46 and 1977–86 (Galbraith, Barnett and Stanwell-Smith, 1987)

Decade	Disease	No. of outbreaks	Decade	Disease	No. of outbreaks
1937–46	Typhoid	3	1977–86	Campylobacter enteritis	5
	Paratyphoid	1		Chemical	1
	Dysentery	1		Viral gastro.	5
	Viral gastro.	2		Other	4
	Total	7		Total	15

The aetiology of waterborne outbreaks in developed nations has changed considerably since the turn of the century (Tables 1.4 and 1.5). For example, typhoid fever has been almost completely eliminated as a cause of waterborne outbreaks. No UK waterborne outbreak of typhoid fever has been reported for almost 30 years. In the USA, the most frequently identified aetiology since 1971 has been giardiasis, while in the UK the most commonly identified cause of waterborne illness is

Table 1.5 Aetiology of waterborne outbreaks in the USA during the periods 1920–40 and 1981–90 (Craun, 1992) (this table should be examined against Table 1.1)

Years	Disease	No. of outbreaks	Years	Disease	No. of outbreaks
1920–40	Typhoid fever	372	1981–90	AGI[*]	128
	Gastroenteritis	144		Giardiasis	71
	Shigellosis	10		Shigellosis	22
	Amoebiasis	2		Chemical	18
	Hepatitis A	1		Viral gastro.	15
	Chemical	1		Hepatitis A	11
				Campylo-bacteriosis	10
	Total	530		Salmonellosis	4
				Cryptosporidiosis	2
				Yersiniosis	2
				Chronic gastro.	2
				E. coli 0157:H7	1
				Typhoid fever	1
				Dermatitis	1
				Cyano-like body	1
				Cholera	1
				Amoebiasis	1
				Total	291

[*]Acute gastro-intestinal illness of unknown aetiology.

due to *Campylobacter* spp. There have been five recorded outbreaks of waterborne Campylobacter enteritis in the UK between 1971 and 1987 (Galbraith, Barnett and Stanwell-Smith, 1987). However, as it has been only possible to identify *Campylobacter* since the mid-1970s, previous outbreaks of unknown aetiology may also have been due to this organism (Dadswell, 1990b). In both the UK and the USA the most frequent aetiology in waterborne outbreaks is viral gastro-enteritis. It is felt that many of these outbreaks represent a combination of viral, bacterial and parasitic aetiologies which present themselves as viral infections, but which cannot be accurately identified at the time of the outbreak (Craun, 1992).

While traditional waterborne diseases such as typhoid and cholera have largely disappeared from Europe and North America, they still remain major killers in poorer developing countries. The World Health Organization calculates that, in Asia and Africa alone, there are up to 5.5 million new cases of cholera with over 120 000 deaths annually (Jones, 1994).

As Tables 1.1, 1.3, 1.4 and 1.5 illustrate, waterborne diseases are now known to be caused by a broader variety of microbes than just enteric bacteria (Sobsey *et al.*, 1993). The growing prevalence of viral and protozoan waterborne diseases presents a significant public health problem for developed as well as developing countries. Viral infections include hepatitis caused by Hepatitis virus types A and E, and gastro-enteritis caused by rotaviruses (section 6.3) (Jones, 1994). It has been reported in the USA that during the period 1976 to 1980, 13 out of a total of 38 serologically confirmed waterborne outbreaks were caused by the Norwalk virus (Singh and McFeters, 1992).

Common protozoans implicated in waterborne disease include *Giardia lamblia*, *Cryptosporidium* spp. and *Entamoeba histolytica*. *Giardia lamblia* has been the most frequently identified aetiologic agent in US waterborne disease outbreaks since 1978 (Herwaldt *et al.*, 1992). Most of these outbreaks have occurred as a result of transmission via untreated surface water or surface water with disinfection as the only form of treatment (Ainsworth, 1990). *Cryptosporidium parvum* was recognized as a waterborne pathogen in 1983, and between 1983 and 1986 some 4000 infections were reported in Britain (Galbraith, Barnett and Stanwell-Smith, 1987). Since 1986 there have been relatively few waterborne outbreaks; however, the majority of those which occurred were due to *Cryptosporidium* sp. By 1990 this organism had become the fourth most commonly identified cause of diarrhoea in the UK (Badenoch, 1990) (section 6.2).

Table 1.6 lists the principal agents responsible for waterborne disease from 1986 to 1990 in the USA. It can be seen that a large number of waterborne disease outbreaks (in some years over 50%) were acute

Table 1.6 The principal causative agents and number of outbreaks of waterborne disease in the USA from 1986 to 1990 (adapted from Herwaldt *et al.*, 1992)

Aetiologic agent	1986–88	1988–90	Total
AGI*	22	14	36
Giardia	9	7	16
Chemical	4	0	4
Shigella	3	0	3
Norwalk-like virus	3	1	4
Salmonella	2	0	2
Campylobacter	1	0	1
Cryptosporidium	1	0	1
CGI†	1	0	1
Hepatitis A	46	2	48
CLB‡ (possible)	0	1	1
E. coli 0157:H7	0	1	1

*Acute gastrointestinal illness of unknown aetiology.
†Chronic gastrointestinal illness.
‡Cyano-like bodies.

gastro-intestinal illness of unknown aetiology. During the period 1989–90, two large outbreaks were caused by *E. coli* 0157:H7 and cyanobacteria-like bodies respectively (section 6.4). These organisms had never previously been associated with waterborne disease (Herwaldt *et al.*, 1992). Since then *E. coli* 0157:H7 has been widely implicated in waterborne outbreaks.

All diseases which may be spread by the faecal–oral route may also be contracted by accidental ingestion of contaminated recreational waters. In addition, outbreaks of *Pseudomonas dermatitis* associated with the use of hot tubs, whirlpool baths and swimming pools have been described (McFeters and Singh, 1992). Other infections associated with warm waters include *Legionella pneumophilia*, *Naegleria fowleria* and *Acanthamoeba*, although infection requires special conditions such as direct inhalation of water or aerosols and so is not strictly due to drinking contaminated water.

1.5   SOURCES OF WATERBORNE DISEASE

INTRODUCTION

Contamination of drinking water by pathogenic and non-pathogenic micro-organisms occurs mainly at source, although contamination can also occur during treatment or within the distribution system. Contamination of an otherwise potable water can also occur within the consumers' premises.

The majority of waterborne disease is attributed to the use of either untreated or inadequately treated ground and surface waters (Table 1.7) (Craun, 1988; Herwaldt *et al.*, 1992). In the USA the use of contaminated, untreated groundwater or inadequately disinfected groundwater was responsible for 44% of waterborne outbreaks reported during 1981–8. In the same period, the use of contaminated, untreated surface waters accounted for 26% of outbreaks (Craun, 1991). Until 1971, more outbreaks were attributed to the use of contaminated, untreated waters than to deficiencies in treatment. Outbreaks in surface water systems today are more the result of inadequate or interrupted disinfection, particularly where chlorination is the only form of treatment (Craun, 1988). In the period 1988–90, 54% of outbreaks were the result of treatment deficiency, with 31% due to untreated waters (Herwaldt *et al.*, 1992). During the period 1981–5 there was a dramatic increase in filtered water systems. As a cause of waterborne outbreaks, the use of contaminated untreated groundwater is now less common than inadequate or interrupted disinfected groundwater. Such incidents arise where little effort has been made to prevent contamination at source and where there is a lack of maintaining continuous disinfection (Craun, 1991)

Galbraith, Barnett and Stanwell-Smith (1987) also found distribution deficiencies and source contamination to be the major causes of waterborne disease in the UK. Of the twenty one cases of public water supply contamination between 1940 and 1987, ten were due to contami-

Table 1.7 Causes of waterborne outbreaks in the USA from 1981 to 1990 (Craun, 1992)

| Cause of outbreak | % of outbreaks by type of water system | | |
	Community	Non-comm.	All systems
Untreated groundwater	12.1	44.3	26.5
Inadequately disinfected groundwater	13.7	33	16.5
Ingestion of contaminated water while swimming	–	–	14.1
Inadequately treated surface water	28.2	9.3	15.1
Distribution deficiencies	24.2	3.1	12.4
Filtration deficiencies	12.9	1	5.8
Unknown	5.6	3.1	3.8
Untreated surface water	1.6	4.1	3.4
Miscellaneous	1.7	2.1	2.4
Total	100	100	100

nation during distribution and eleven were due to source contamination. Eight of these eleven were due to defective or no chlorination.

In recent years, an increasing number of outbreaks due to contamination by protozoan and viral agents have been observed. Many of these organisms are more resistant to physico-chemical treatments, including disinfection, than bacterial pathogens (Sobsey *et al.*, 1993). Most outbreaks of *Giardia* occur in untreated waters or waters where the only form of treatment is chlorination (Ainsworth, 1990). The first reported outbreaks of cryptosporidiosis occurred in Cobham in Surrey (UK) in 1983 and at Braun Station, Texas in 1984. The only form of water treatment in these cases was chlorination, which the organism could easily by-pass (Barer and Wright, 1990; Gray, 1994). The levels of chlorine which would be required to destroy protozoan oocysts are far in excess of those levels that can be allowed in water supplies (Ainsworth, 1990). Enteric viruses have also been shown to cause outbreaks of disease in waters considered suitable for consumption. A massive outbreak of hepatitis in Delhi in 1958 was later proven to be the result of inadequate chlorination (Dennis, 1959).

ANIMALS AND BIRDS

Wild and domestic animals and birds in the watershed are a considerable reservoir of intestinal pathogens and have been implicated as causative agents of source water contamination (Craun, 1988). Because of their aquatic habits, beavers and muskrats have a high potential for contaminating waterways. A US survey in Washington found that over a three year period, the prevalence of *Giardia* infection ranged from 6–19% in beavers and from 35–43% in muskrats (Akin and Jakubowski, 1986). Microbiologists in Libya have recorded contamination at a water irrigation plant to be the result of droppings from a large population of birds (Jones, 1994). Seasonal habitation of wildlife on lakes may introduce *Salmonella*, *Giardia*, *Campylobacter* or other waterborne pathogens (Geldreich, 1992; Tauxe, 1992). While migratory waterfowl which roost at reservoirs have no serious deleterious effect on water quality, gulls which feed on contaminated and faecal material at refuse tips and sewage works have been shown to excrete pathogens. All five British species of *Larus* gulls have been rapidly increasing in recent years with the Herring gull (*Larus argentatus*) doubling its population size every 5–6 years. This has resulted in a very large increase in the inland population of *Larus* gulls throughout the British Isles, both permanent and those over-wintering (Gould, 1977). These birds are opportunist feeders and have taken advantage of the increase in both the human population and its standard of living, feeding on contaminated waste during the day and then roosting on inland water bodies, including reservoirs, at night (Hickling, 1977). Faecal bacteria, especially *Salmonella* spp., have been traced from feeding sites such as domestic waste-tips to the reservoir, showing that gulls are directly responsible for the dissemination of bacteria and other human pathogens (Benton *et al.*, 1983; Riley *et al.*, 1981). Many reservoirs have shown a serious deterioration in bacterial quality due to contamination by roosting gulls and those situated in upland areas where the water is of a high quality, so that treatment is minimal before being supplied to the consumer, are particularly at risk from contamination (Benton *et al.*, 1983; Fennell *et al.*, 1974; Jones, 1978). Numerous *Salmonella* serotypes can be isolated from gull droppings as well as faecal coliforms, faecal streptococci and spores of *Clostridium welchii* (Gould, 1977). Ova of parasitic worms have also been isolated from gull droppings and birds are thought to be a major cause of the contamination of agricultural land with the eggs of the human beef tapeworm (*Taenia saginata*) (Crewe, 1967). There is a clear relationship between the number of roosting gulls on reservoirs and the concentration of all the indicator bacteria. In pristine areas with little or no human population, animals are more likely to be the primary source of waterway contamination (Akin and Jakubowski, 1986). Hence, the potential exists for contamination of all watersheds and surface waters by infected animals and birds as their complete exclusion cannot be guaranteed (Craun, 1988).

WATER DISTRIBUTION

Water which is microbiologically pure when it enters the distribution system may undergo deterioration before it reaches the consumer's tap. Contamination by micro-organisms can occur through air valves, hydrants, booster pumps, service reservoirs, cross connections, back syphonage or through unsatisfactory repairs to plumbing installations. Further problems can also arise within the domestic plumbing system.

The main danger associated with drinking water is the possibility of it becoming contaminated during distribution by human or animal faeces. This was the case in the Bristol outbreak of giardiasis in 1985, when contamination occurred through a fractured main (Browning and Ives, 1987). A major outbreak of typhoid fever occurred in 1963 in Switzerland when sewage seeped into the water mains through an undetected leak in the pipe. There are many more examples indicating that the quality of microbial water is potentially at risk while in the distribution system. Uncovered service reservoirs can be a major source of contamination, especially from birds. Normally the main is under considerable positive pressure; however, back syphonage can occur in the distribution system if the water pressure drops and there are faulty connections or fractures in the pipe. In this way contaminants can be sucked into the distribution system. The problems are more likely to occur when water supply pipes are laid alongside sewerage systems.

On passing through the distribution system, the microbiological properties of the water will change. This is due to the growth of microorganisms on the walls of the pipes and in the bottom sediment and debris. While not normally causing any serious health problems, these non-pathogenic bacteria can cause serious quality problems, causing discoloration and deterioration in taste and odour. Increased heterotrophic bacteria within the distribution system are due to a number of factors, usually the absence of a residual disinfectant combined with either contamination from outside the distribution network or, more commonly, from regrowth. The growth of bacteria both in the water and on pipe surfaces is limited by the concentration of essential nutrients in the water. Organic carbon is the limiting nutrient for bacterial growth, with the growth of bacteria being directly related to the amount of assimilable organic carbon (AOC) in the water (Block, 1992).

Among the major genera found in distribution systems are *Acinetobacter*, *Aeromonas*, *Listeria*, *Flavobacterium*, *Mycobacterium*, *Pseudomonas* and *Plesiomonas* (sections 5.7 and 6.4). Some of these organisms can be considered as opportunistic pathogens. The type of microorganisms and the numbers depend on numerous factors such as water source, type of treatment, residual disinfectant, nutrient levels in treated water and so on. The development of slimes, or biofilms as they are known, lead to the survival of other bacteria. Certainly biofilms protect bacteria from disinfectants. *Legionella* in particular is able to survive within biofilms as are *Pseudomonas* and *Aeromonas* spp. (Stout, Yu and Best, 1985). Development of non-pathogenic coliforms is possible in biofilms, but when detected it is important for the water operator not to assume a non-faecal cause such as regrowth in the distribution system, even in the absence of *E. coli*.

With good treatment and adequate disinfection with chlorine ensuring a residual disinfection to control regrowth within the distribution system and to deal with any contamination, then serious problems should not arise. When supplies have a high chlorine demand, due to the presence of organic matter and humic acids for example, then it is very difficult to maintain sufficient residual in the system. Also it is necessary to strike a balance so that those closest to the treatment plant on the distribution network do not receive too large a dose of chlorine in their water and those at the end too little. Recent research has shown that many bacteria, viruses and protozoan cysts are far more resistant to chlorination than the indicator bacteria normally used to test disinfection efficiency. So new guidelines are being prepared to optimize disinfection of water supplies. Currently the residual concentration of free chlorine leaving the treatment plant is below 1.0 mg/L and normally nearer to 0.5 mg/L. Most treated waters continue to exert some chlorine demand which reduces this free chlorine residual even further. Further chlorine is lost by its interaction with deposits and corrosion products within the distribution system, so that free chlorine residuals do not often persist far into the network, leaving the water and the consumer at risk (Mathieu *et al.*, 1994; Paquin *et al.*, 1992).

The problem stems from treatment processes being primarily designed to remove particulate matter rather than dissolved organic matter. Both chlorination and ozonation lead to increases in AOC concentration in water, while coagulation and sedimentation can remove up to 80% of the AOC present. Assimilable organic carbon is also removed by biological activity during filtration, although residence times are far too short to permit significant reductions. Success has been achieved by chemically oxidizing the AOC into simpler compounds to make them more readily removable by biological filtration. Methods to determine the growth potential of heterotrophic bacteria in drinking water distribution systems has been reviewed by Huck (1990) and van der Kooij (1990). There is evidence that the micro-organisms living on granular activated carbon filters can also significantly reduce AOC concentrations (Huck, Fedovak and Anderson, 1991).

The presence of biofilms is currently causing water microbiologists much concern, especially as pathogenic micro-organisms can be protected from inactivation by disinfectants by the biofilm. Control of these growths is difficult and research is very active in this area. However, the risk of chlorination by-products which are known to be carcinogenic at higher concentrations than those found in water must be balanced with the risk of microbial pathogens from inadequate disinfection. The problems of biofilms are reviewed by Wende, Characklies and Smith (1989) and Maul, Vagost and Black (1991).

HOUSEHOLD PLUMBING SYSTEMS

Back syphonage in plumbing systems is more of a problem in older buildings as modern building regulations and water bylaws incorporate measures to prevent it. It generally occurs when a rising main supplying more than one floor suffers a loss of pressure at a low point in the system, causing a partial vacuum in the rising main. Atmospheric pressure on the surface of, for example, a bath full of water on an upper floor in which a hose extension or a shower attachment from an open tap has been left, will push the contents of the bath back up through the hose and tap into the plumbing system in order to fill the partial vacuum. Plumbing systems suffer from constant changes in pressure, so care should always be taken when hoses are left to run water into containers. This is extremely important if private supplies are pumped by a submersible pumping system from an underground storage tank. Once the pump is switched off any water left in the pipework connected directly to the pump will syphon back into the tank. If the pump has been used to fill a watering can or a water trough, for example, if the end of the hose is not removed before the pump is switched off then the contents of the container will be drained by back syphonage into the storage tank. Mechanical backflow prevention devices are available to prevent water flowing back in to the household plumbing system or even the mains distribution system from the consumers' premises (White and Mays, 1989).

Water storage tanks must be covered. When left open they become a breeding ground for a wide variety of micro-organisms. They also become fouled by birds and vermin. Many houses will have mice in the attic, a favourite place for over-wintering, under or in the inches of insulation. The only source of water in the attic is the cold water storage tank, so fouling by vermin is fairly common. In 1984 there was an outbreak of *Shigella sonnei* at a boarding school in County Dublin, which caused widespread gastro-intestinal illness amongst the inmates. The source was identified as the water storage tank which was supplying all their drinking water and which had been fouled by pigeons.

Legionnaires' disease first came to notice in 1976 when there was a major epidemic among the residents of a hotel in Philadelphia. It was at this hotel that the American Legion was holding its annual conference, from which the disease takes its name. Since then numerous outbreaks of the disease have been reported including many in the UK. In 1985, 75 000 cases were reported in the USA with 11 250 deaths attributed to the disease which is a severe form of pneumonia. Although there are over 20 strains of *Legionella pneumophila* known, only one serotype is thought to be a serious threat to health. The bacterium has been associated with domestic water systems, especially hot water which is stored between 20 °C and 50 °C. Heat exchangers, condensers in air conditioning units, cooling towers and shower heads have all been found to be havens for the bacteria leading to human infection. Legionella are commonly found in hospital water systems, and hospital-acquired (nosocomial) Legionnaires' disease is now a major health problem (Marrie *et al.*, 1994; Patterson *et al.*, 1994). Between 1980 and 1992 there were 218 confirmed cases of nosocomial Legionnaires' disease with 68 deaths reported in England and Wales. During that period, 2347 cases were reported of which 60% acquired their infection in the UK, associating hospitals with 15% of all cases (Joseph *et al.*, 1994). It appears that a long retention time, and the presence of key nutrients such as iron, provides ideal conditions for the bacteria to develop. Therefore iron storage cisterns subject to corrosion will be susceptible. In addition, water pipes that are installed alongside hot water pipes or other sources of heat may also permit the bacteria to develop. Infection is by inhalation of contaminated aerosols from contaminated appliances. The bacteria survive and grow within phago-cytic cells, multiplying in the lungs causing bronchopneumonia and tissue damage. The bacterium is widespread in natural waters so any water supply can become contaminated (States *et al.*, 1990). Prevention of *Legionella* infection is normally achieved by either hyperchlorination or thermal irradiation of infected pipework, although these methods are not always successful or the effect permanent. The only effective preventative method has been shown to be the use of UV sterilizers as close to the point of use as possible (Liu *et al.*, 1995).

Non-pathogenic bacteria develop in rubber extension pipes on taps and in mixer taps, leading to taste and odour problems. Long rubber extension pipes which extend into contaminated water (during dish or clothes washing at the kitchen sink) will quickly develop thick microbial contamination inside the pipe. Such pipes should be avoided whenever possible because they can cause a significant reduction in water quality, and can also result in back syphonage. Micro-organisms will also grow on activated carbon granules held in cartridges used in home water treatment systems, and may, ironically, cause aesthetic as well as health problems. So if consumers are not changing the cartridges regularly, then it is best not use them at all (Gray, 1994).

1.6   RISK ASSESSMENT

Any discussion on drinking water quality standards must include some consideration of risk assessment. Put in simplest of terms, risk assess-ment is an attempt to quantify the possible health consequences of human exposure to particular circumstances (Contruvo, 1989). Risk to human health is defined as the likelihood or probability that a given exposure to a microbial pathogen may induce illness or damage to health of the exposed individuals. Risk assessment involves the analysis of past waterborne outbreaks and the adverse health effects, and the prediction of the likely consequences of future exposures (Philipp, 1991). All human activities carry some degree of risk, with exposure to risk from the consumption of food and water inevitable and largely unavoidable (Pochin, 1975). Greene (1982) states clearly that society gains immense benefits from water and faces certain risks in the process.

In drawing up any standard pertaining to public safety, a certain balance must be achieved between the benefits of safety and the costs of achieving it. When considering the risks presented by consumption of drinking water the following must be carefully evaluated:

- What is the current health risk posed by modern day water supplies to public health?
- Is this an acceptable risk?
- What is the cost of upholding stringent standards in monetary and social terms?

In considering the risks to human health from consumption of water supplies a number of factors must be taken into account:

- Identification and characterization of the micro-organisms causing illness by the water route.
- The limits of analytical tools for pathogen and indicator detection.
- Occurrence and distribution of the organism.
- Dose-response relations, especially for those populations most at risk (Sobsey *et al.*, 1993).

This information can then be used to determine the risks involved in terms of human health and social costs.

Rose and Gerba (1991) have established the probability of infection from exposure to one organism and the minimum dose to obtain a 1% infection within the community for all the major waterborne pathogens (Table 1.8). Moynihan (1992) argues that to get these figures it is assumed that (1) the organisms are randomly distributed, which she feels is unlikely for aggregates of microbes; (2) that the consumer population is equally susceptible to a single organism exposure, which again is unlikely, and (3) that the exposure is defined as the consumption of two litres of water per day. Taking these limitations into account, these results show that a risk of 1% infection within a community is 10 to 100 times higher from a protozoan or viral pathogen than for bacteria. This work is particularly exciting in that it is possible for the first time to predict the possible infection rate within a community once the concentration of pathogens in finished water is known. Rose and Gerba give an example of a community exposed to concentrations of poliovirus and *Giardia* cysts at a contamination level of 0.1 to 100 organisms per 100 litres. Assuming a daily intake of drinking water of 2 litres per day, then 0.3 to 300 infections of poliovirus and 0.4 to 400 infections of *Giardia* per 10 000 of the population can be expected.

It is quite likely that the majority of the cases shown in Table 1.8 were benign with no long-lasting effects. Rather than denigrating the significance of waterborne disease, such statistics serve to illustrate the success of current standards and the use of indicator organisms to date. When compared with the potential risks associated with exposure to

Table 1.8 Risk of infection from exposure to one organism and minimum doses required to obtain a 1% infection within a community for a range of pathogens (adapted from Rose and Gerba, 1991)

Micro-organism	Risk of infection from exposure to a single organism	Minimum dose (no. of organisms) required for 1% infection
Campylobacter	7×10^{-3}	1.4
Salmonella	2.3×10^{-3}	4.3
Salmonella typhi	3.8×10^{-5}	263
Shigella	1.0×10^{-3}	10
Vibrio cholera	7×10^{-6}	1428
Poliovirus 1	1.49×10^{-2}	0.67
Poliovirus 3	3.1×10^{-2}	0.32
Echovirus 1,2	1.7×10^{-2}	0.59
Rotavirus	3.1×10^{-1}	0.03
Entamoeba coli	9.1×10^{-2}	0.1
Entamoeba histolytica	2.8×10^{-1}	0.04
Giardia lamblia	1.98×10^{-2}	0.5

Infection was the end result measured for each organism, except *Shigella dysenteriae*. It is assumed that the population is equally susceptible to a single exposure, the latter being defined as the consumption of 2 litres of water per day.

chemicals in water, actual health risks associated with microbes are high (Table 1.9). Also, while the risk of infection of some degree is a certainty for most people, thanks to modern medicine the actual morbidity and mortality hazard is relatively small for those in developed countries. However, because those within the population most at risk are the young and the weak, even a low level of risk is not easily tolerated by consumers (Greene, 1982). An additional factor in the evaluation of the risk present in drinking water is the increased prevalence of immuno-compromised and other high-risk individuals within the population. This has made the evaluation of risks associated with consumption of contaminated waters even more difficult particularly given that the potential virulence of conditional pathogens present in the heterotrophic plate count (Chapter 5) for such individuals has yet to be determined (Sobsey *et al.*, 1993). The risk for those in developing countries, where there is often poor water treatment and management as well as inadequate medical support, is very much greater, with many millions of people, mainly children, dying from waterborne diseases annually.

Drinking water standards are developed from guidelines which describe the quality of water suitable for consumption under all circumstances, taking into account environmental, social, economic and cultural factors (Moore, 1989). An interesting study in the UK examined

Table 1.9 Comparison of outbreaks and cases of waterborne disease with chemical poisoning in Scotland from 1948 to 1987 (adapted from Benton et al., 1989).

Disease	Percentage of total outbreaks	Percentage of total cases
Typhoid fever	7	0.4
Paratyphoid fever	7	0.16
Salmonellosis	2	0.2
Shigellosis	5	30.1
Campylo-bacteriosis	7	0.4
Viral gastro-enteritis	5	52.2
Giardiasis	2	0.03
Total pathogen illness	35	83.49
Chemical poisoning	37	0.71
Unknown	28	15.8

a number of outbreaks of giardiasis in Avon and Somerset (Gray, Gunnell and Peters, 1994), and attempted to identify the key risk factors associated with the disease. Such studies are prerequisites to quantifying risks to consumers and recreational users of water. Ideally, an adequately treated water should not contain any risks from biological contaminants; however, this is an unattainable and unrealistic goal (Cabelli, 1978). Because bacteria can be generally effectively removed from drinking water supplies, the recommended guidelines for total coliforms and E. coli is often set at zero (EC, 1980; Federal Register, 1988, 1989; WHO, 1993). This approach has on the whole proved to be largely satisfactory. However, the idea of risk associated with standards needs further clarification; for example, when faecal coliforms are found in litre samples of water, but not in the normal 100 mL volumes tested (Moynihan, 1992). For microbial waterborne disease to occur the pathogen must be present in sufficient concentrations to cause infection and the development of the disease. Also a susceptible host has to come into contact with the pathogen at that minimum critical dose. The basis for the use of coliforms as indicator organisms is that they react to water treatment and disinfection in much the same way as other pathogens. This is untrue even for the primary bacterial pathogens such as Salmonella and Shigella spp. that are more resistant to certain disinfectants than faecal coliforms (Payment, 1991; Sobsey, 1989). Viral and protozoan pathogens do not behave like bacterial indicators and the realization of this has placed a serious question mark over their continued use in establishing safety of drinking waters.

Establishing water quality criteria for viruses is difficult as the minimum infective dose for all strains is generally unknown. A guideline value of zero has been suggested but has met with considerable opposition given the present lack of epidemiological evidence and the potential costs that implementing such legislation would involve. Setting standards for protozoans is much easier in that there is more information available on infective doses (Table 1.8). It should be emphasized that both viruses and protozoan pathogens are notoriously resistant to disinfection and the emphasis is towards source protection rather than removal or disinfection during treatment (Moore, 1989).

A major concern, in relation to the establishment of standards for drinking water quality, is that regulations can place undue emphasis on benefits, thereby generating excessive costs for the consumer. This issue has generated much debate in the USA and developing nations in recent years (US EPA, 1990a).

1.7   LEGISLATION GOVERNING MICROBIAL WATER QUALITY

EU LEGISLATION

The EU Directive relating to the quality of water intended for human consumption, or the Drinking Water Directive as it is more commonly known, was implemented in July 1980 (EC, 1980). This Directive applies to all water supplied for human consumption within the European Community and also that used in processing food products intended for human consumption. The water quality criteria on which the Directive is based were drawn largely from the WHO guidelines for drinking water quality (WHO, 1984), although significant differences exist between the WHO guideline values and those in the Directive itself (Crowley and Packham, 1993; Gray, 1994).

The Directive lists 66 parameters which are classified into six groups: (A) Organoleptic, (B) Physicochemical, (C) Substances undesirable in excessive amounts, (D) Toxic substances, (E) Microbiological and (F) Minimum concentrations for softened water.

The EU normally sets two standards for parameters in the form of concentrations not to be exceeded. The G value is the guide value which the Commission desires Member States to work towards in the long term. The I value is the mandatory value in Directives and is the minimum standard that can be adopted by Member States. In the Drinking Water Directive, the I value is called the maximum admissible concentration or MAC. National standards must conform to the MAC values, although it is permissible for Member States to set stricter standards, which does happen occasionally. The Directive sets out G or

MAC values for some 60 compounds, with MAC values only set for 44 of these parameters, although guidance is given for the others.

The EU Drinking Water Directive sets a number of microbiological parameters using indicator organisms. The following MAC values apply to finished waters:

- Zero total coliforms per 100 mL sample. Where a sufficient number of samples are examined then a 95% consistent result is acceptable.
- Zero faecal coliforms (*E. coli*) per 100 mL sample.
- Zero faecal streptococci per 100 mL sample.
- Less than one for sulphite reducing clostridia per 20 mL sample.
- There should be no significant increase in the total bacteria colony counts above background levels, although a guide value (G) of <10 per mL at 22 °C and <100 per mL at 37 °C has been set. Denmark has recently set MAC values of <20 and <200 per mL at 22 and 37 °C respectively.

Analysis is to be done by membrane filtration for total coliforms, faecal coliforms and faecal streptococci, and by the multiple tube method for sulphite reducing clostridia. The EU Directive does not require either viruses or protozoan pathogens (cysts or oocysts) to be routinely measured, although it makes it clear that water intended for human consumption should not contain any pathogenic organisms. However, in terms of regulated parameters it relies totally on the indicator organisms listed above. The Directive does contain the rather ambiguous statement: 'If it is necessary to supplement the microbiological analysis of water intended for human consumption, the samples should be examined not only for the bacteria listed but also for pathogens including salmonella, pathogenic staphylococci, faecal bacteriophages, and entero-viruses; nor should such water contain parasites, algas or other organisms such as animalcules'. The Directive also specifies minimum sampling frequencies for coliforms and colony counts (Hayes, 1989) (Table 1.10).

Certain exceptions from the parameters specified are permitted due to climatic or geographical factors. However, such exceptions are not extended to microbiological parameters.

European implementation of the Directive has been much slower than anticipated. Originally, the Directive was required to be implemented by Member States by August 1982. However, the implementation period has in reality been much longer due mainly to problems associated with the interpretation and incorporation into existing regulations (Fawell and Miller, 1992; Crowley and Packham, 1993). Table 1.11 lists some of these problems, although all EU member states have now incorporated the Drinking Water Directive into law, with France being the last to do so (Gray, 1994). In Ireland the Directive was

Table 1.10 Sampling frequency at consumers' taps in the UK, as required by the Water Supply (Water Quality) Regulations for microbial parameters (HMSO, 1989)

Parameters	Water supply zone		Sampling frequency (number per year)		
	Volume distributed for domestic purposes (m³/d)	Population supplied	Reduced	Standard	Increased
			Ground- Surface water water		
Total coliforms	≤ 100	≤ 500	12		
Faecal coliforms	101–1 000	501– 5 000	12		
Residual disinfectants	1 001–2 000	5 001–10 000	24		
Colony counts	2 001–4 000	10 001–20 000	48		
		20 001–50 000	*		

*At the rate of 12 per 5000 population except in the case of colony counts where the standard number is 52.

enacted in 1988 by the publication of the European Communities (Quality of Water Intended for Human Consumption) Regulations 1988 (Flanagan, 1990), while in England and Wales implementation occurred with the enactment of the Water Supply (Water Quality) regulations 1989, as amended by the Water Supply (Water Quality) (Amendment) Regulations 1989 and 1991. Both the UK and Ireland have set their microbial drinking water standards at the MAC levels.

Table 1.12 gives a rough guide to the implementation of the Drinking Water Directive in the EU to date. It is important to bear in mind that comprehensive statistics on drinking water are not available throughout

Table 1.11 Problems associated with implementation of the Drinking Water Directive (Crowley and Packham, 1993)

Problem	Comment
1. Absolute nature of MAC values.	
2. Scientific practicality.	The Directive gives no scientific justification for any of the standards.
3. Priority for improvement.	No guidance given on the relative priority to be attached to meeting all the Directive standards. There is no distinction made between those parameters of public health and those of aesthetic importance.
4. Monitoring.	Unrealistic.
5. Guide values.	Status uncertain.

Table 1.12 Implementation of the Drinking Water Directive in Europe (from Fawell and Miller, 1992)

Country	Legal instrument	First effective date
Belgium	Royal Decree 1984, State circular 1988, Regional law 1989	April 1984
Denmark	Statutory orders 1980, 1983, 1988	April 1980
France	Public Health Code and Decree 1961, Decree 1988, Decree 1989	January 1989
Germany (West)	Federal Law on Epidemics, Law on Foodstuffs and Consumer Goods, Drinking Water Regulations, amended 1986, 1990	October 1986
Greece	Ministerial Decrees, Social Service on the Quality of Drinking Water 1968, 1974, 1986	January 1986
Ireland	Statutory Instrument No. 81 European Communities [Quality of Water for Human Consumption] Reg. 1988	January 1988
Italy	Prime Minister's Decree 1985, Presidential Decree 1988	May 1988
Luxembourg	Royal Decree 1985	April 1985
Netherlands	Water Works Act 1957, Statutory Instruments 1960, Amendments 1984	July 1984
Portugal	Water Decree Law 1990	1990
Spain	Royal Decrees 1982, 1990	June 1982
UK	Administrative Circular 1982, Water Supply [Water Quality] Regulations 1989, 1990, 1991	September 1989, England and Wales; May 1990 Scotland; Not implemented Northern Ireland

the EU. This makes comparisons difficult except on a piecemeal basis. It is very difficult to assess the level of compliance within various Member States as many countries have decentralized monitoring systems and compliance data is generally not readily available. The annual report of the Drinking Water Inspectorate (Department of the Environment, 1994b) showed that in 1993 in England and Wales water companies demonstrated compliance with the relevant water quality standards for 98.9% of the time. Coliforms were not detected in the finished water produced by 81.5% (1351) of the companies' 1657 water treatment works. In 98.5% of the 5068 service reservoirs, coliforms were absent from at least 95% of all samples (Table 1.13).

In May 1995 the newly revised Drinking Water Directive was published in its draft form (EU, 1995). In many ways it is a simpler document proposing 3 microbial, 26 chemical and 16 indicator parameters. As well as complying with new fixed parameters, water

Table 1.13 Summary of the microbiological quality of water leaving treatment works in England and Wales (Department of the Environment, 1994b; 1995)

	1994	1993	1992	1991
No. of water treatment works	1603	1657	1683	1717
Coliforms				
Total no. of determinants	218 770	225 705	222 947	215 190
No. containing coliforms	336	438	586	734
% containing coliforms	0.2	0.2	0.3	0.3
Treatment work with coliforms detected	242	306	385	445
% of all works	15	18	23	26
Faecal coliforms				
Total no. of determinants	218 784	255 728	222 937	214 318
No. containing faecal coliforms	83	135	158	231
% containing faecal coliforms	<0.1	<0.1	<0.1	<0.1
Treatment works with faecal coliforms	72	117	127	173
% of all works	4	7	8	10

supplied under the Directive will have to be free of pathogenic micro-organisms and parasites in numbers constituting a danger to public health, although as before there are no guide values given for specific viruses, protozoans or bacterial pathogens. The proposed Directive gives separate maximum permissible concentrations for the microbial parameters for both tap and bottled waters, except natural mineral waters which have their own Directive. For tap water maximum permissible concentrations are given for *E. coli* (0 per 100 mL), faecal streptococci (0 per 100 mL), and sulphite reducing clostridia (0 per 20 mL). For bottled waters the maximum permissible concentrations are stricter with *E. coli* (0 per 250 mL), faecal streptococci (0 per 250 mL) and sulphite reducing clostridia (0 per 50 mL). *Pseudomonas aeruginosa* has also been included for the first time with a maximum permissible concentration of 0 per 250 mL. The microbial parameters are listed in a new section, Part A, of the Directive. Part B contains chemical parameters and Part C indicator parameters. Included in Part C are total coliforms with a maximum permissible concentration of 0 per 100 mL for tap waters and 0 per 250 mL for bottled waters; and also included are total bacterial counts which must not show any abnormal change. Specific and detailed notes are given on analysis, including the composition of all recommended media.

The recommended analyses in the proposed Directive are:

- **Total coliforms:** membrane filtration followed by incubation on membrane lauryl broth for four hours at 30 °C followed

by 14 hours at 37 °C. All yellow colonies are counted regardless of size.

- *E. coli*: membrane filtration followed by incubation on membrane
  lauryl broth for four hours at 30 °C followed by 14 hours at 44 °C. All
  yellow colonies are counted regardless of size.
- **Faecal streptococci:** membrane filtration followed by incubation on
  membrane enterococcus agar for 48 hours at 37 °C. All pink, red or
  maroon colonies which are smooth and convex are counted.
- **Sulphite reducing clostridia:** maintain the sample at 75 °C for 10
  minutes prior to membrane filtration. Incubate on tryptose-sulphite-
  cycloserine agar at 37 °C under anaerobic conditions. Count all black
  colonies after 24 and 48 hours incubation.
- *Pseudomonas aeruginosa*: membrane filtration followed by incubation
  in a closed container at 37 °C on modified Kings A broth for 48
  hours. Count all colonies which contain green, blue or reddish-brown
  pigment and those that fluoresce.
- **Total bacteria counts:** incubation in a yeast extract agar for 72 hours
  at 22 °C and for 24 hours at 37 °C. All colonies to be counted.

The proposed Directive is expected to be adopted during 1996. Apart
from parameter changes it also contains a number of new provisions,
for example the use of check and audit monitoring to confirm analysis
and reporting, and the revision of parameter values in light of scientific
and technological advances every 10 years.

US LEGISLATION

Legislation to protect bacteriological quality of drinking water in the
USA began with the publishing of the Public Health Service Act of
1914. Initially these regulations were not mandatory for all public water
supplies (Crowley and Packham, 1993). This situation changed in 1974
with the passing of the Safe Drinking Water Act (SWDA). This act
required the US EPA to establish national standards for drinking water
quality. Federal responsibility was now extended to include all public
water systems serving 25 or more persons or from which 15 service
connections are taken (Masters, 1991). Interim regulations were
published in 1975 (US EPA, 1976). These regulations provided for a
minimum number of samples to be examined each month and established the maximum number of coliforms (Maximum Containment
Level or MCL) allowable per 100 mL of finished water. The SWDA was
amended in 1986, significantly expanding the original Act. Under this
Amendment:

- The number of contaminants to be examined was extended to 83.
- The equivalent of filtration was required for all water systems.

- Disinfection was required for all water systems.
- The use of lead in any pipe, solder, flux or fittings in any public water system was prohibited.
- Enforcement procedures were streamlined which included raising penalties to $25 000 per day for infringements (US EPA, 1990b).

The amended Act also mandated the US EPA to specifically regulate for *Giardia*, viruses, *Legionella*, heterotrophic bacteria and turbidity, in addition to coliform counts (US EPA, 1990b, 1993). A Maximum Containment Level Goal (MCLG) and a Maximum Containment Level are issued for various parameters. MCLs are enforceable standards which apply to parameters considered to have an adverse effect on human health and are set as close to the MCLG as is technologically and financially feasible. Maximum Containment Level Goals are non-enforceable health goals which are set at levels at which no known or anticipated adverse health effects will occur and which allow an adequate margin of safety (US EPA, 1990a).

The standards are split into Primary and Secondary Drinking Water Standards. The Primary National Drinking Water Standards cover those parameters which are considered to be harmful to health and include coliforms. The Secondary National Drinking Water Standards are more concerned with aesthetic values such as taste and colour.

In June 1989, the US EPA promulgated a revised total coliform rule (Federal Register, 1989) which became effective on 31 December 1990. The main elements of this revised rule are as follows:

- MCLs are based on the presence or absence of coliforms in a sample. For systems analysing at least 40 samples per month no more than 5% of the monthly samples may be coliform positive. The MCLs for those systems collecting less than 40 samples per month is no more than one coliform positive sample per month.
- Each public water system must take samples in accordance with a written sampling plan.
- The number of routine samples is based on the population served by the system.
- Procedures are outlined in the event of a positive sample. If the system is found to be in violation of the MCL for total coliforms then the public is notified.
- Certain conditions are established whereby a total coliform positive result may be invalidated.

A 100 mL sample bottle must be used in analysing total coliforms using one of the following techniques: 10-tube multiple tube fermentation technique, the MF technique, the P–A coliform test or the minimal-media ONPG-MUG test (Borup, 1992; Berger, 1992) (Chapter 4).

This presence–absence concept has a number of potential advantages:

- Sensitivity is improved because it is more accurate to detect coliform presence than to make quantitative determinations.
- The concept is not affected by changes in coliform density during storage.
- Data manipulation is much improved (Moore, 1989).

The US EPA regulations are developed through a visible and extensive consultation process. Proposed new regulations are opened for public comment which must be considered by the EPA before a final rule can be made. This makes for a very lengthy decision-making process (Crowley and Packham, 1993).

WHO RECOMMENDATIONS

The WHO guidelines for drinking water are perhaps the most important standards relating to water quality standards. They are used universally and are the basis for both EU and US legislation. The original guidelines were published in two volumes in 1984. Volume 1 is the guidelines while volume 2 contains the scientific evidence on which the recommendations in volume 1 are based. The existing guidelines were based on the available toxicological evidence up to 1981, and so are very much out of date. The revision began in 1987 with new guidelines finally agreed in Geneva in September 1992. The new guidelines include microbiological, chemical and radiological parameters. The chemical parameters include 17 inorganics, 27 organics, 33 pesticides and 17 disinfectants and associated by-products. The revised guidelines for microbial parameters are shown in Table 1.14.

There are no specific guideline values for either viruses or parasites. This is because it is considered that the analytical methods for these organisms are too costly, complex and time consuming for routine laboratory use. Instead, guideline criteria are outlined based on the likely viral content of source waters and the degree of treatment necessary to ensure that even large volumes of water have a negligible risk of containing viruses. It is considered that 'the attainment of the bacteriological criteria and the application of treatment for virological reduction should ensure the water presents a negligible health risk'. These revised guidelines should have a significant effect on existing standards and have in fact led to new proposed EU drinking water standards (EU, 1995).

MONITORING REQUIREMENTS

Under the EU Drinking Water Directive, parameters to be monitored are grouped under four headings which reflect the frequencies with

Table 1.14 World Health Organization drinking water guide values for bacteriological quality of drinking water (WHO, 1993)

Organisms	Guideline
All water intended for drinking	
E. coli or thermotolerant coliform bacteria[*†]	Must not be detectable in any 100 mL sample
Treated water entering the distribution system	
E. coli or thermotolerant coliform bacteria[*]	Must not be detectable in any 100 mL sample
Total coliform bacteria	Must not be detectable in any 100 mL sample
Treated water in the distribution system	
E. coli or thermotolerant coliform bacteria[*]	Must not be detectable in any 100 mL sample
Total coliform bacteria	Must not be detectable in any 100 mL sample. In the case of large supplies where sufficient samples are examined, must not be present in 95% of samples taken throughout any 12-month period

Immediate investigative action must be taken if either *E. coli* or total coliform bacteria are detected. The minimal action in the case of total coliform bacteria is repeat sampling; if these bacteria are detected in the repeat sample, the cause must be determined by immediate further investigation.

[*]Although *E. coli* is the more precise indicator of faecal pollution, the count of thermotolerant coliform bacteria is an acceptable alternative. If necessary, proper confirmatory tests must be carried out. Total coliform bacteria are not acceptable indicators of the sanitary quality of rural water supplies, particularly in tropical areas where many bacteria of no sanitary significance occur in almost all untreated supplies.
[†]It is recognized that in most rural water supplies in developing countries faecal contamination is widespread. Under these conditions, the national surveillance agency should set medium-term targets for the progressive improvement of water supplies.

which they are to be examined. These groups are: minimum monitoring (C1), current monitoring (C2), periodic monitoring (C3) and occasional monitoring for special situations or in the case of accidents (C4). The principle underlying the EU monitoring strategy is that the extent to which sampling and analysis is carried out increases with the magnitude of population being served by a given supply (Flanagan, 1990). While the Drinking Water Directive itself lays down minimum requirements for monitoring, much is left to the discretion of the competent national authorities (EC, 1980).

In England and Wales, statutory responsibility for monitoring water quality is placed upon the water companies and is subject to checks by the Local Authorities and by the Drinking Water Inspectorate. The basic

Table 1.15 Minimum frequency of sampling and analysis for tap and bottled waters (EU, 1995)

1. Minimum frequency of sampling and analyses
(Except for water offered for sale in bottles or containers)

Volume of water distributed or produced each day within a supply zone (m³)		Check monitoring number of samples per year	Audit monitoring numbers of samples per year
	⩽ 100	1	1
> 100	⩽ 1 000	1	1
> 1 000	⩽ 2 000	3	1
> 2 000	⩽ 10 000	12	1
> 10 000	⩽ 20 000	60	1
> 20 000	⩽ 30 000	120	2
> 30 000	⩽ 60 000	180	3
> 60 000	⩽ 100 000	365	6
> 100 000	⩽ 200 000	730	10
> 200 000	⩽ 300 000	1 460	20

2. Minimum frequency of sampling and analysis for water offered for sale
in bottles or containers (provisional)

Volume of water produced for offering for sale in bottles or containers each day* (m³)		Check monitoring number of samples per year	Audit monitoring number of samples per year
	⩽ 1	1	1
> 1	⩽ 10	1	1
> 10	⩽ 20	3	1
> 20	⩽ 100	12	1
> 100	⩽ 200	60	1
> 200	⩽ 300	120	2
> 300	⩽ 600	180	3
> 600	⩽ 1 000	365	6
> 1 000	⩽ 2 000	730	10
> 2 000	⩽ 3 000	1 460	20

*The volumes are calculated as averages taken over a calendar year.

unit for monitoring purposes is the water supply zone, an area designated by a water company in which no more than 50 000 people live. A discrete area served by a single source will always be delineated as a single supply zone unless it serves a population in excess of 50 000, in which case it is divided into two or more water supply zones. Water companies are required to take a specified standard number of samples for each parameter from each water supply zone per year. The number of samples is dependent on the population being served (HMSO, 1989).

For microbiological parameters, at least 50% of the sampling points must be random (Table 1.10). New sampling frequencies have been proposed in the revised Drinking Water Directive (EU, 1995). Separate tables for minimum frequency of sampling and analysis are given for tap and bottled waters (Table 1.15).

In the USA, the Revised Coliform Rule requires a monitoring

Table 1.16 Number of water samples to be taken, under the US EPA Drinking Water Regulations, for total coliform analysis according to the population served by the water supply system (US EPA, 1990b)

Population served	Minimum no. of routine samples per month*	Population served	Minimum no. of routine samples per month
25 to 1 000[†]	1[‡]	59 001 to 70 000	70
1 001 to 2 500	2	70 001 to 83 000	80
2 501 to 3 300	3	83 001 to 96 000	90
3 301 to 4 100	4	96 001 to 130 000	100
4 101 to 4 900	5	130 001 to 220 000	120
4 901 to 5 800	6	220 001 to 320 000	150
5 801 to 6 700	7	320 001 to 450 000	180
6 701 to 7 600	8	450 001 to 600 000	210
7 601 to 8 500	9	600 001 to 780 000	240
8 501 to 12 900	10	780 001 to 970 000	270
12 901 to 17 200	15	970 001 to 1 230 000	300
17 201 to 21 500	20	1 230 001 to 1 520 000	330
21 501 to 25 000	25	1 520 001 to 1 850 000	360
25 001 to 33 000	30	1 850 001 to 2 270 000	390
33 001 to 41 000	40	2 270 001 to 3 020 000	420
41 001 to 50 000	50	3 020 001 to 3 960 000	450
50 001 to 59 000	60	3 960 001 or more	480

*In lieu of the frequency specified, a Non-Community Water System (NCWS) using groundwater and serving 1000 persons or fewer may monitor at a lesser frequency specified by the state until a sanitary survey is conducted and reviewed by the state. Thereafter, NCWSs using groundwater and serving 1000 persons or fewer must monitor in each calendar quarter during which the system provides water to the public, unless the state determines that some other frequency is more appropriate and notifies the system (in writing). Five years after promulgation, NCWSs using groundwater and serving 1000 persons or fewer must monitor at least once/year.

A NCWS using surface water, or groundwater under the direct influence of surface water, regardless of the number of persons served, must monitor at the same frequency as a like-sized Community Water System (CWS). A NCWS using groundwater and serving more than 1000 persons during any month must monitor at the same frequency as a like-sized CWS, except that the state may reduce the monitoring frequency for any month the system serves 1000 persons or fewer.

[†]Includes public water systems which have at least 15 service connections, but serve fewer than 25 persons.

[‡]For a CWS serving 25–1000 persons, the state may reduce this sampling frequency, if a sanitary survey conducted in the last five years indicates that the water system is supplied solely by a protected groundwater source and is free of sanitary defects. However, in no case may the state reduce the frequency to less than once/quarter.

program based not on an estimation of coliform density, but on the presence or absence in a sample. Three levels of monitoring are required by this rule. The first level of sampling is the monthly routine sample. The number of samples taken is again based on the population being served (Borup, 1992) (Table 1.16). In the event of a coliform positive routine sample, a set of repeat samples must be collected. Finally if any repeat or routine samples are totally coliform positive, the sample is then analysed to determine the presence of faecal coliforms or E. coli. Statistical analysis of the new rule by Borup (1992) indicates that water of acceptable quality may be found in violation of the standard, and equally waters of unacceptable quality may be found to meet standards, especially when small numbers of samples are taken. This is due to the randomness in monitoring.

In terms of priorities, maximum efforts in a monitoring programme should be devoted to distribution systems in public water supplies because all of the population served by the system is assumed to drink the water and be at some risk at all times (Geldreich and Kennedy, 1982). Most waterborne outbreaks now occur in small community and non-community systems (Craun, 1992). As monitoring programmes in general tend to be dependent on populations served, violations in smaller systems tends to be missed. Analysis of data by Geldreich and Kennedy (1982) suggest that for smaller water supply units increased sampling frequency would provide more intensive monitoring and represent a considerable saving in the long term. However, for non-community systems, where such an approach would be impractical, a more effective monitoring programme has yet to be developed.

Monitoring strategies are intended to establish baseline data against which microbiological data can be compared in the event of a water-borne outbreak or other unusual circumstances relating to water quality. According to Geldreich (1992), current monitoring strategies are not entirely realistic in that they tend not to take into account unpredictable quality changes such as storms, which may have adverse effects on treatment barriers. Future monitoring of surface waters should not be solely confined to daily samples, but should be tailored to account for the intensity of major storm events over a watershed which may result in contamination.

1.8 CONCLUSIONS

Developments throughout the twentieth century in the area of water quality control have resulted in the virtual eradication of waterborne disease in developed countries. This success is not reflected in developing countries where it is estimated that 80% of rural populations of Asia, Africa and South America do not have access to safe waters, and where primary bacterial pathogens still present a major public health problem.

Chlorination as a form of water treatment has made a major contribution to the reduction of waterborne disease particularly in more affluent nations, where there has been a significant drop in the number of outbreaks following its introduction. However, since the 1940s, there has been a slight but consistent increase in the incidence of waterborne disease. This increase may be attributable to better reporting or more sensitive detection methods.

Recent years have seen the emergence of new threats to water quality. These are chiefly protozoan and viral agents which appear to be more resistant to water treatment processes than bacteria. The key unit processes in conventional water treatment were developed decades ago and were designed to cope with classic bacterial waterborne diseases. It is clear that these traditional unit processes (e.g. rapid sand filtration, chlorination) are not adequate barriers to remove or destroy these newly emerging pathogens and that new technology and better operational control are required.

The incorporation of microbiological parameters into standards worldwide is a recognition of their importance in relation to maintaining water quality. However, these standards differ widely in the way in which they are interpreted and subsequently implemented because of state mandates, individual countries' regulations and, more frequently, the finances available for water quality monitoring.

Indicator organisms and the coliform concept

2

2.1 THE ORIGINS OF INDICATOR ORGANISMS

Once the connection had been made between outbreaks of diseases such as cholera and typhoid fever, and contaminated water, attempts were made to develop methods which would isolate the causative organisms. In practice, however, this proved to be very difficult for a number of reasons. Pathogenic organisms such as *Vibrio cholerae* and *Salmonella* spp. are usually present in very low numbers in water (requiring several litres of water to be examined) and tend to die away very quickly (requiring repeated sampling and rapid analysis). Their isolation and detection can be very difficult, often requiring specific and expensive techniques which are generally considered impractical for routine testing (Department of the Environment, 1994a; Department of the Environment, Department of Health and Social Security, and Public Health Laboratory Service, 1983). In addition, the wide variety of pathogens that exists presents limitations to the usefulness of procedures for their enumeration (Pipes, 1982b). It has to be emphasized that the ability to detect pathogenic organisms is certainly possible and is steadily being improved with new developments. However, the isolation of pathogens can never be the sole procedure used in routine water analysis, because, as already mentioned, their appearance is intermittent, of short duration, and the organisms are readily attenuated and few (Bonde, 1977). There also exists the likelihood that the water will have been consumed by the time the pathogen is detected. The revised manual for bacterial examination of drinking water supplies (Department of the Environment, 1994a) warns that the impression of security given by microbiological testing of water at infrequent intervals may be quite false. It stresses that the value of microbial testing is dependent upon test frequency and regular use. It emphasizes in the introduction of the manual in bold type that 'it is far more important to examine a [drinking water] supply frequently by a simple test than occasionally by a more complicated test or series of tests'.

In an attempt to find a more practical alternative to the isolation of specific pathogens microbiologists, as long ago as 1885, began to look to other organisms whose presence might indicate the possibility of faecal contamination, i.e. an indicator organism. The underlying assumption of this approach is that water intended for consumption should contain none of these pathogens in the first place (WHO, 1993).

CRITERIA FOR INDICATOR ORGANISMS

To be of value as an assessor of faecal contamination, indicator organisms should satisfy the following criteria:

- They should be a member of the normal intestinal flora of healthy people.
- They should be exclusively intestinal in habitat and therefore exclusively faecal in origin if found outside the intestine.
- Ideally they should only be found in humans.
- They should be present when faecal pathogens are present and only when faecal pathogens are expected to be present.
- They should be present in greater numbers than the pathogen they are intended to indicate.
- They should be unable to grow outside the intestine with a die-off rate slightly less than the pathogenic organism.
- They should be resistant to natural environmental conditions and to water and wastewater treatment processes in a manner equal to or greater than the pathogens of interest.
- They should be easy to isolate, identify and enumerate.
- They should be non-pathogenic (Feacham et al., 1983; Oliveri, 1982).

The concentration of indicator organisms should be related to the extent of faecal contamination and by implication to the concentration of pathogens and the incidence of waterborne disease (Pipes, 1982a). There is no absolute correlation between the numbers of the indicator organism present and the actual presence or numbers of enteric pathogens. The finding of an indicator organism in a properly treated water indicates the presence of material of faecal origin and thus the potential of contamination.

In practice, there is no organism or group of organisms which meets all of the above requirements; however, many of them are fulfilled by the coliform group. For nearly a century this group has been the principal means by which the sanitary quality of water has been determined. Several other organisms also meet many of these requirements and are widely used to provide supplementary information in certain circumstances, principally faecal streptococci and *Clostridium perfringens*. In Europe the Drinking Water Directive specifies numerical standards for total and faecal coliforms, faecal streptococci, sulphite-reducing clostridia and total viable bacterial counts at 22 °C and 37 °C (EC, 1980). There are other possible indicator organisms and chemicals, and these are discussed in detail in Chapter 5.

2.2 THE DEVELOPMENT OF COLIFORMS AS INDICATOR ORGANISMS

The earliest methods of water bacteriology were aimed at direct isolation of specific pathogenic organisms. The cultivation of *Cholera vibrio* from water and faecal samples was largely successful at that time, although difficulties were encountered with the detection of typhoid-like organisms (Bonde, 1977). Bacteriologists attempting to overcome the difficulty of isolating such organisms found that human faeces contained large numbers of aerobic, Gram-negative organisms that resembled typhoid bacteria. These bacteria were subsequently found to be bile salt-tolerant, facultatively anaerobic organisms capable of growth and of fermenting lactose at 37 °C. Among this group, later known as the coliforms, was *Bacterium coli*. This organism was isolated from human faeces in 1885 by Escherich (Dadswell, 1990a). The coliforms were quickly recognized as possible indicators of faecal pollution and procedures were developed to detect and estimate their numbers in raw waters. Further advances in bacteriological techniques and the development of more suitable media meant that even small numbers of coliforms could be detected. The most widely adopted method of isolation was inoculation of MacConkey Broth and incubation at 37 °C for 48 hours. Cultures producing acid and gas were subsequently subcultured onto MacConkey Agar to eliminate obligate anaerobes. By definition a coliform became any organism that was isolated by this method.

As bacteriology advanced there were increasing reports of recoveries of coliform organisms from non-faecally contaminated environments, although *Bacterium coli* was not isolated in such waters. The presence of *B. coli* was realized to be a definite indicator of faecal contamination (Dadswell, 1990a). These observations stimulated research into separating faecal from non-faecal coliforms. In 1904, Eijkman found that faecal coliforms could produce gas from glucose at 47 °C, while coliform strains from non-faecal sources failed to grow at these temperatures (Wolf, 1972), although it was subsequently found that some non-faecal coliforms could also grow at these elevated temperatures, particularly from tropical areas (Dadswell, 1990a). This problem was resolved in the 1930s by McKenzie who demonstrated that in addition to growth at 44.5 °C (the most acceptable temperature for separation of faecal from non-faecal coliforms), *B. coli* type I, now renamed *Escherichia coli*, was also capable of growth and of fermenting lactose at this temperature with the production of gas, and producing indole from tryptophan. These properties still remain the criteria by which *E. coli* is separated from total coliforms.

DEFINITION OF THE COLIFORM GROUP

The coliform group consists of several genera of bacteria belonging to the family Enterobacteriaceae. Traditionally these genera included *Escherichia, Citrobacter, Enterobacter* and *Klebsiella*. However, using more modern taxonomical criteria, the group is heterogeneous and includes non-faecal lactose fermenting bacteria as well as other species which are rarely found in faeces but are capable of multiplication in water (WHO, 1993) (Table 2.1).

Historically, the definition of the coliform group has been based on methods used for its detection rather than on the tenets of systematic bacteriology (APHA, 1992). Accordingly, when the multiple tube method is used, the American Public Health Association defines

Table 2.1 Nomenclature of Enterobacteriaceae associated with natural waters (APHA, 1992)

Tribe	Genus	Species
Escherichieae	*Escherichia*	*E. coli*˙
	Shigella	*S. dysenteriae, S. flexneri, S. boydii, S. sonnei*
Edwardsielleae	*Edwardsiella*	*E. tarda*
Salmonelleae	*Salmonella*	*S. cholerae-sius, S. typhi S. enteritidis*
	Arizona	*A. hinshawii*˙
	Citrobacter	*C. freuneii,*˙ *C. diversus*˙
Klebsielleae	*Klebsiella*	*K. pneumoniae K. ozaenae, K. rhinoscleromatis,*˙ *K. oxytoca*˙
	Enterobacter	*E. cloacae,*˙ *E. aerogenes,*˙ *E. agglomerans*
	Serratia	*S. marcescens, S. liquefaciens, S. rubidaea*
	Hafnia	*H. alvei*˙
Proteeae	*Proteus*	*P. vulgaris, P. mirabilis*
	Providencia	*P. alcalifaciens, P. stuartii, P. rettgeri*
Erwinieae	*Morganella*	*M. morganii*
	Erwinia Pectobacterium	(Amylovara, Herbicola, Caratova groups)
Yersinieae	*Yersinia*	*Y. pestis, Y. enterocolitica, Y. pseudotuberculosis, Y. ruckeri, Y. intermedia, Y. fredericksenii*

˙Species that may ferment lactose with gas.

coliforms as 'all aerobic and facultatively anaerobic, Gram negative, non spore-forming, rod shaped bacteria that ferment lactose with acid and gas production'. Where the membrane filtration technique is used, the coliform group is normally defined as comprising all aerobic and facultatively anaerobic, gram-negative, nonspore-forming, rod-shaped bacteria that develop a red colony with a metallic sheen within 24 hours at 35 °C on an Endo type medium containing lactose (APHA, 1992). The WHO definition is broader and refers to gram-negative, rod shaped bacteria capable of growth in the presence of bile salts or other surface active agents with similar growth inhibiting properties, able to ferment lactose at 35–7 °C with the production of acid, gas and aldehyde within 27–48 hours. They are also oxidase negative, non spore-forming and display ß-galactosidase activity. With the development of methods for detecting coliforms which do not rely on characteristics such as the production of acid and gas from lactose, the use of method-related definitions become largely obsolete and a more scientific definition is required.

A particular problem lies with the fermentation of lactose by the enzyme ß-galactosidase into glucose and galactose. Clearly the possession of the gene coding for the production of ß-galactosidase is the most fundamental characteristic of the Enterobacteriaceae (coliforms). Yet the production of gas from lactose has been found to be extremely variable. The expression of this gene can be affected by many factors including time, temperature and the growth medium used, so that the same organism may or may not ferment lactose sufficiently to register as a lactose fermenter under different test conditions (Department of the Environment, 1994a). It is now widely accepted that any new definition of a coliform must be based on the possession of the ß-galactosidase gene.

The coliform group also includes the thermotolerant faecal coliforms. These are defined as being able to ferment lactose at 44 °C (WHO, 1993), and not only include *E. coli* but also species of the *Klebsiella*, *Enterobacter* and *Citrobacter* genera. *Escherichia coli* is considered to be the only true faecal coliform as other thermotolerant coliforms can be derived from non-faecally contaminated waters.

The coliform index initially involves examining the sample for total coliforms. Total coliforms are largely faecal in origin, but also include species which are commonly found in unpolluted soils and vegetation, and therefore do not present a public health problem. Subsequently, total coliform results are interpreted as presumptive results. The sample is then examined for thermotolerant coliforms (Department of the Environment, Department of Health and Social Security, and Public Health Laboratory Service, 1983). Although this group is often termed faecal coliforms, this is not entirely correct, as some non-faecal organisms are also capable of growth at 44 °C, such as non-faecal *Klebsiella* spp. The growth of such organisms may result in high counts that may be interpreted as faecal coliforms (Bayley and Seidler, 1977). In the EU Drinking Water Directive the term 'faecal coliform' is used specifically to indicate coliforms of faecal origin which it defines as those that are thermotolerant, i.e. capable of growth at 44 °C. As not all thermotolerant coliforms are faecal in origin, they must be regarded as presumptive faecal coliforms (Department of the Environment, 1994a). Therefore the presence of *E. coli*, which is known to be exclusively faecal in origin, is usually also determined. *Escherichia coli* consists of up to 95% of the enterobacteria found in faeces (Waite, 1985). In addition to the production of acid and gas at 44 °C, *E. coli* is also able to produce indole from tryptophan and most strains produce ß-glucuronidase (Department of the Environment, 1994a; Department of the Environment, Department of Health and Social Security, and Public Health Laboratory Service, 1983). Generally for assessment of the microbiological quality of surface waters, it is the faecal coliform (*E. coli*) count which is primarily determined, due to its public health implications. However, for treated drinking waters enumeration of total coliforms is generally sufficient since it is assumed that waters designated for human consumption should not contain any micro-organisms (Cabelli, 1978). The isolation of total and faecal coliforms (*E. coli*) is considered in detail in Chapter 3.

2.3 COLIFORMS AS INDICATORS OF WATER QUALITY

The coliform index is still widely considered the most reliable indicator for potable water (Pipes, 1982b). The fact that the index remains the main microbiological parameter in most water quality standards is an acknowledgement of this (APHA, 1992; Department of the Environment, Department of Health and Social Security, and Public Health Laboratory Service, 1983; WHO, 1993). Its retention in the new Drinking Water Directive will ensure that it remains the key indicator of microbiological quality of drinking water throughout Europe until the Directive is next revised in the year 2007. However, for many years a growing number of workers from a wide variety of disciplines have questioned the use of coliforms as a measure of water quality. It is considered that the coliform concept was developed based 'on decisions and assumptions which were largely correct in the light of knowledge available at the time' (Waite, 1985). Since its adoption, tremendous advances have been made in public health microbiology, but many of the developments made in research have not been applied to practical use. This has inevitably resulted in the maintenance of practices which are perhaps no longer relevant to present day public health situations.

Those critical of the use of coliforms in water quality assessment have identified several deficiencies of the group as indicator organisms:

- regrowth in aquatic environments;
- regrowth in distribution systems;
- suppression by high background bacterial growth;
- not indicative of a health threat;
- a lack of correlation between coliforms and pathogen numbers;
- no relationship between protozoan and viral numbers;
- the occurrence of false positive and false negative results.

Each of these problems is dealt with below.

COLIFORM REGROWTH IN THE AQUATIC ENVIRONMENT

Ideally an indicator organism should not be able to proliferate to a greater extent than enteric pathogens in the aquatic environment (Feacham *et al.*, 1983). Figure 2.1 shows the results of a study by Dutka (1973), demonstrating that coliforms are able to grow and multiply readily in aquatic environments as opposed to faecal streptococci which rarely multiply in water (Geldreich, 1970).

This ability of coliforms to regrow in surface waters, also known as aftergrowth, is not only limited to *Enterobacter*, *Klebsiella* and *Citrobacter* (Ellis, 1989), but also *E. coli* (Roszak and Colwell, 1987). The die-off rate of indicator bacteria in surface waters depends on a number of factors that include the quality of the receiving waters and its temperature. Mancini (1978) has suggested that 90% of the coliform population in fresh water dies off within 120 hours at 0 °C but within 15 hours at 30 °C. Shuval, Cohen and Kolodney (1973) observed that coliforms and total coliforms were capable of regrowth even in chlorinated sewage (Figure 2.2). They found that regrowth occurred even when coliforms were not detectable in the final water after chlorination. Other studies have shown that coliforms are able to reproduce in enriched waters and thus falsely indicate an elevated health hazard (Geldreich, 1970; Dutka, 1973; Dutka, 1979; Pipes, 1982b). High coliform counts have been reported in waters receiving pulp and paper mill effluents, sugar beet wastes and raw domestic sewage (Geldreich, 1970; Dutka, 1973; Pipes, 1982b). Modern agricultural practices have meant that large quantities of nitrogenous fertilizers are often released into natural waters which may subsequently lead to algal blooms. As a result of autolysis these algae release their cell contents into the water. In either situation the elevated levels of nutrients present retards or reverses bacterial dieaway and as a result, regrowth and extended persistence of coliforms and faecal coliforms may occur (Fox, Keller and van Schothoist, 1988). In Australia, Asbolt, Dorsch and Banens (1995) have observed annual

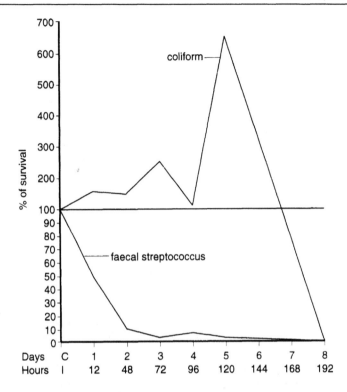

Figure 2.1 Study of the survival and multiplication of coliforms and faecal streptococci in relatively unpolluted lake waters (Dutka, 1973).

natural blooms of faecal coliforms in Sydney Water Corporation's largest raw water reservoir. These blooms have been found to be almost completely composed of E. *coli* and occur in such densities that they cannot be accounted for by external factors such as surface runoff or discharges, nor can they be related to rainfall, algal blooms or other limnological factors.

A significant observation has been made with regard to coliform survival and growth in marine environments. *Escherichia coli*, when subject to the stress of the marine environment, can assume a quiescent form which is undetectable by normal detection procedures (Fox, Keller and van Schothoist, 1988).

COLIFORM REGROWTH IN DISTRIBUTION SYSTEMS

Research in aftergrowth of bacteria in distribution systems can be divided into (1) investigations of bacteria isolated in drinking waters and (2) those isolated from distribution system wall or pipe surfaces

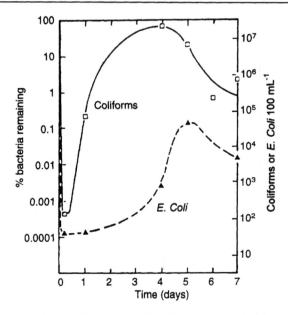

Figure 2.2 Regrowth of coliforms and *E. coli* in sewage effluent after inactivation with 5 mg/litre chlorine (Shuval, Cohen and Kolodney, 1973).

(Olson and Nagy, 1984). It has been well established that bacterial numbers in drinking waters increase as they pass through the distribution system (Olson and Nagy, 1984). Such growth may be the result of a variety of different situations within the distribution system, for example:

- Loss or lack of residual disinfection. Growth of heterotrophic bacteria and coliforms may occur in distribution system sections that are unable to maintain an effective disinfectant. This is especially common in spur mains, particularly those which are very long, as found in rural areas, and/or are serving only a few houses.
- Inadequate treatment may result in a breakthrough indicated by the occurrence of coliforms or high heterotrophic bacteria in finished waters or effluents.
- Injured coliforms may recover within the distribution system.
- Particle-associated bacteria may break through treatment barriers (Le Chevallier, 1990).

Numerous researchers have shown that increased resistance to disinfection may result from attachment or association of micro-organisms to various surfaces including macro-invertebrates (Crustacea, Nematoda, Platyhelminthes and Insecta), algae, small carbon particles and colloidal or suspended solids (Reasoner, 1988). Le Chevallier, Evans and Seidler (1981) observed that high turbidity can provide a mechanism for coliform survival. If the organisms are embedded in suspended particles, then chlorine is not able to come in contact with them. In addition, turbidity interferes with coliform detection by the membrane filtration technique and subsequently particle-associated organisms may go undetected.

The ability to survive and grow on the surface of drinking water distribution pipes (or biofilm growth) has been relatively well established since the turn of the century (Olson and Nagy, 1984). There is ample microscopic evidence available which shows that most distribution pipe surfaces are colonized by micro-organisms (Le Chevallier, Babcock and Lee, 1987), with pipe surfaces colonized by large populations of bacteria and diatoms. In contrast, Ridgeway and Olson (1981) found the pipe surfaces they examined to be sparsely populated with organisms. Despite the differences, these observations indicate that bacterial growth is a normal occurrence on distribution pipe surfaces (Le Chevallier, 1990). The growth of bacteria on pipe surfaces is limited by the concentration of essential nutrients in the water, in particular organic carbon. The growth of bacteria can be directly related to the availability of assimilable organic carbon in the water. This has already been considered in detail in section 1.5.

Most water systems have encountered consumer complaints resulting from microbial activity within the distribution network. However, the type of problem-causing organism generally found (i.e. iron, sulphate and manganese bacteria) results in corrosion and water discoloration affecting the taste, odour and appearance of the water, which, although not aesthetically pleasing, do not present public health problems (Gray, 1994). What does present problems, however, is when regrowth of coliforms and faecal coliforms occur within the distribution network. In recent years both seasonal and continuous occurrences of coliform organisms (e.g. *Klebsiella*, *Enterobacter* and *Citrobacter* spp.) have been reported in water systems (Le Chevallier, Schulz and Lee, 1991; Smith, Hess and Hubbs, 1989; Wierenga, 1985).

Le Chevallier (1990) identifies several factors which distinguish chronic coliform growth in distribution network. These are:

- No coliforms detected in treatment plant effluents even when the most sensitive methodologies are employed, indicating that the treatment plant is not the main source of coliform bacteria.
- High coliform densities are routinely detected in distribution system samples despite the fact that very low numbers are entering the system from the treatment plant. For example in a study of distribution system biofilms, Le Chevallier, Babcock and Lee (1987) showed

that even though finished treatment effluents contained low coliform levels (0.3 cfu/100 mL), coliform densities increased twentyfold (0.64 cfu/100 mL) as the water moved from the treatment plant through the distribution system.
- The reoccurrence of coliforms which lasts for a prolonged length of time (years). This can be largely attributed to the difficulty inactivating biofilm bacteria.

These investigations showed that coliforms found in drinking water samples originated from distribution system biofilms. Coliform bacteria can coexist with chlorine residuals under certain circumstances (Geldreich, 1996). *Escherichia coli* is 2400 times more resistant to chlorine when attached to a surface than as free cells in water, leading to high survival rates within water distribution systems (Le Chevallier, Cawthon and Lee, 1988a,b). Further studies indicated that in addition to a low level of injured or stressed coliforms, coliform diversity increased as the water flowed through the distribution system. These points would indicate that provided the physico-chemical characteristics of the water and distribution systems are suitable, the distribution system is quite hospitable to coliform growth and reproduction. The health significance of coliform growth in distribution systems is an important consideration for water utilities, since the presence of these bacteria may mask the presence of indicator bacteria in water supplies resulting from a true breakdown of treatment barriers.

Biofilms in distribution systems have proved difficult to inactivate and control. Several investigators have shown that chlorine levels used in water treatment are inadequate to inactivate biofilms. Le Chevallier, Cawthon and Lee (1988b) have demonstrated that a free chlorine residual level as high as 4.3 mg/L is inefficient in eliminating coliform occurrence, while Ridgeway *et al.* (1984) found that a residual of 15–20 mg chlorine per litre was required to control biofilm fouling of reverse osmosis membranes. Chlorine has a very distinctive odour and a taste threshold of only 0.16 mg/L at pH 7 rising to 0.45 mg/L at pH 9. So while a mild chlorine odour in drinking water is generally acceptable to consumers as a sign that the water is microbially safe, concentrations in excess of the taste threshold makes water most objectionable and is generally rejected by the consumer. Therefore it is clear that routine chlorination practices, which rely on a residual effect, will not prevent or control bacterial regrowth. It has been suggested that monochloramine might be more effective for biofilm control than chlorine (Table 2.2). While monochloramine has a longer-lasting disinfecting effect than chlorine, it is less effective than free chlorine (Gray, 1994). The use of other disinfectants or combination of disinfectants with biological treatment has also been explored (Le Chevallier, 1990).

Table 2.2   Inactivation of *Klebsiella pneumoniae* attached to glass microscopic slides (Le Chevallier, Cawthon and Lee, 1988b)

C × T multiple*	% Reduction of viable count ± standard deviation	
	Free chlorine	Monochloramine
1.0		91.5 ± 5
1.5		91.5 ± 6
2.0		94.1 ± 5
2.5		96.2 ± 6.5
3.0		>99.9
75	34 ± 16.9	
100	51.3 ± 18.2	
125	91.1 ± 6	
150	99.6 ± 0.4	

*C × T values – milligram minutes per litre. For unattached *K. pneumoniae* inactivation occurred at C × T ratings of 0.065 mg min/litre for free chlorine and 33 mg min/litre for monochloramine.

SUPPRESSION OF COLIFORM GROWTH BY HIGH BACKGROUND BACTERIAL GROWTH

A number of bacterial species are known to be antagonistic to coliform growth. These antagonists include strains of *Pseudomonas*, *Sarcina*, *Micrococcus*, *Flavobacterium*, *Bacillus*, and *Actinomyces* as well as some yeasts (Hutchinson, Weaver and Scherago, 1943). Le Chevallier, Seidler and Evans (1980) observed that chlorinated waters containing high numbers of antagonists had low coliform isolation rates. Conversely when the antagonist population was low (less than 20% of the total plate count), a greater coliform incidence (75%) was recorded (Table 2.3). It is thought that such suppression can result in an underestimation of coliform numbers by as much as 80% (Hutchinson, Weaver and Scherago, 1943; Oliveri, 1982). The microbial quality of water supplies in distribution systems is fully discussed by Geldreich (1996).

Table 2.3   Relationship between percentage of coliform antagonists and the presence of coliforms (Le Chevallier, Seidler and Evans, 1980)

Sample	No.	No. with coliforms	Occurrence (%)
Distribution			
>20%	16	3	19
<20%	7	4	57
Raw water			
>20%	0	0	–
<20%	11	11	100

COLIFORMS ARE NOT INDICATIVE OF A HEALTH THREAT

Figures 2.3 and 2.4 illustrate the theoretical relationship between coliforms, faecal coliforms and *Salmonella* sp. However, this relationship seldom seems to hold in actual waterborne disease situations (Olson and Nagy, 1984). A comparative study of community and non-community water systems by Craun, Batik and Pipes (1983) showed no statistical differences between systems in which an outbreak had occurred and systems in which no outbreak had been reported. In fact they found that it is possible to find coliforms in systems for which there are no reported outbreaks and to have outbreaks in systems for which there are no positive coliform results (Table 2.4).

In a study of coliform species in water systems reporting the presence of coliforms, Geldreich and Rice (1987) found a number of incidences where a non-faecal *Klebsiella* was the prominent coliform. This species is

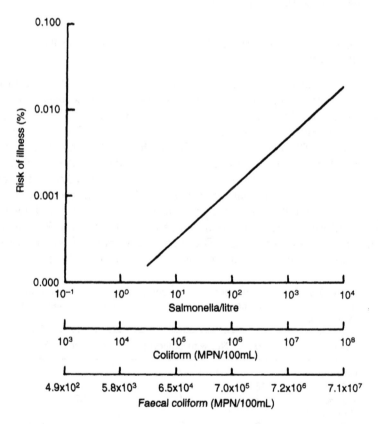

Figure 2.3 Relationship between disease risk and salmonella, coliforms and faecal coliforms (Olson and Nagy, 1984).

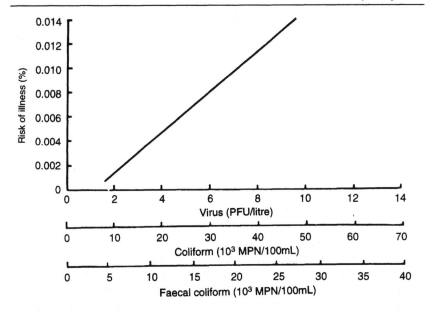

Figure 2.4 Relationship between disease risk and viruses, coliforms and faecal coliforms (Olson and Nagy, 1984).

able to multiply rapidly and can colonize the natural environment (Knittel *et al.*, 1977). In other cases, the reported coliform was either *Citrobacter* or *Enterobacter*. In none of the reported cases were either *E. coli* or faecal coliforms detected nor were any coliforms found positive by the faecal coliform test. Further investigations of these systems revealed that no waterborne disease outbreak had occurred. It was concluded that the incidence of coliforms was not due to faecal contamination but reflected colonization within the distribution system, as discussed earlier. Such incidences can result in repeated and unnecessary orders to boil water before drinking and serve no purpose except to undermine public confidence in municipal water systems. A more

Table 2.4 Non-community water systems: comparison of coliform monitoring results three months prior to outbreak (Craun, Batik and Pipes, 1983)

	Coliform results		
	Positive result	Negative result	Total
Non-community system experiencing an outbreak	8	8	16
Non-community system not experiencing an outbreak	343	455	798
Total	351	463	814

recent study by Payment, Franco and Siemiatycki (1993) shows that a significant proportion of the endemic level of gastro-intestinal illness was attributable to drinking water that was free of known pathogens and met current water quality standards. Again, colonization within the distribution system was considered to be the causative factor.

THERE IS NO CORRELATION BETWEEN COLIFORM AND PATHOGEN NUMBERS

Various studies have investigated the numerical relationship that exists between coliforms and pathogens, with some authors supporting (Bonde, 1977; Morinigo et al., 1990) and others rejecting (Carter et al., 1987; Dutka and Bell, 1973; Hood, Mois and Blake, 1983; Kaper et al., 1979) the hypothesis of a positive correlation between both variables. While it recognized that there is no absolute correlation between coliforms and bacterial pathogens due to the spatially variable and essentially unpredictable behaviour of pathogens (Bonde, 1977; Geldreich, 1970; Townsend, 1992), the underlying principle of the coliform index is that its presence in waters indicates the potential presence of pathogens. There have been reports of where Vibrio sp. (Kaper et al., 1979) and Salmonella sp. (Dutka and Bell, 1973; Morinigo et al., 1990) have been recovered from waters containing few or no coliforms or faecal coliforms. This may be due to a number of factors, for example coliforms have a faster die off rate than Salmonella sp. (Borrego et al., 1990); also, Salmonella typhi has been reported to be more resistant to chlorination than coliforms (Dutka, 1973). In any case, the lack of reliability of coliforms to indicate the presence of bacterial pathogens has prompted some investigators to suggest that coliform and faecal coliform tests be supplemented or replaced with direct detection of the pathogen (Dutka and Bell, 1973; Morinigo et al., 1993; Townsend, 1992). Such action could subsequently render the use of indicators for bacterial pathogens obsolete.

THERE IS NO RELATIONSHIP BETWEEN PARASITIC PROTOZOANS AND VIRUSES

In recent years parasitic protozoans and viruses have replaced bacterial pathogens as the primary causative agents of waterborne disease. In the USA Giardia lamblia has been the most frequently identified agent in waterborne disease outbreaks since 1988 (Herwaldt et al., 1992), while in the UK Cryptosporidium is the fourth most common cause of water-borne-related diarrhoea (Badenoch, 1990).

Many outbreaks attributed to protozoan or viral agents have been reported in conventionally treated water supplies, or in waters which have been minimally treated, all of which met coliform standards (Hejkal *et al.*, 1982). It is now accepted that coliform bacteria are not reflective of the concentration of enteric viruses in natural waters (Geldenhuys and Pretorius, 1989; Gerba *et al.*, 1979; La Belle *et al.*, 1980; Lucena *et al.*, 1982; Metcalf, 1978). Viruses can persist longer at lower temperatures and remain infectious for many months at temperatures near freezing. They have been shown to persist longer in natural waters than faecal coliforms and are more resistant to water and wastewater treatment processes (Nasser, Tchorch and Fattal, 1993). A similar situation exists for protozoan cysts which have also been shown to be more resistant to chlorination and other forms of water treatment than coliforms (Ainsworth, 1990). Analysis of data produced by Rose, Darbin and Gerba (1988) showed no association between coliform bacteria and either *Cryptosporidium* oocysts or *Giardia* cysts (Table 2.5).

Of greatest concern in this instance were occasional peaks of oocysts or cysts where concurrent coliform levels were zero. These findings repeatedly suggest the inadequacy of coliforms as indicators for these particular pathogens, which is particularly significant given their increasing occurrence, and therefore the need for a sensitive indicator.

Table 2.5 Correlation coefficients for coliform bacteria, turbidity and protozoa in a watershed (Rose, Darbin and Gerba, 1988)

	Total coliforms	Faecal coliforms	*Cryptosporidium*	*Giardia*
Turbidity	0.277	0.288	0.242	0.284
Total coliforms		0.709[‡]	0.154	0.018
Faecal coliforms			0.291	0.102
Cryptosporidium				0.778[‡]

Levels of significance $p < 0.05$[*], $p < 0.01$[†], $p < 0.001$[‡]

THE SIGNIFICANCE OF FALSE POSITIVE AND FALSE NEGATIVE RESULTS

The problem of false positive and false negative results are considered two of the more limiting deficiencies of coliforms as faecal indicators. This subject has been reviewed in considerable depth by Waite (1985) and Bonde (1977).

False positive results

False positive presumptive reactions have been attributed to a number of causes:

- the presence of aerobic or facultatively anaerobic, spore-forming, gas-producing bacteria;
- synergistic fermentation of lactose by gas-producing symbionts;
- seasonal occurrence of coliform organisms not surviving the detecting procedures employed;
- microbial antagonism and the presence of oxidase-positive bacteria capable of producing gas from lactose (e.g. *Aeromonas* spp.) (Hussong *et al.*, 1981). Observations by Mates and Shaffer (1989) have shown that when using the membrane filtration technique, oxidase positive organisms will be counted as coliforms (Table 2.6) and in some instances can represent up to 14% of coliforms counted.

Certain species, in particular the aeromonads, are able to mimic the Enterobacteriaceae (Waite, 1985), resulting in inflated total coliform counts. These organisms can produce acid and gas at 37 °C and are therefore regarded as presumptive coliforms unless excluded by further confirmatory tests. The interference of coliform counts by the growth of aeromonads is widely recognized (Grabow and Du Preez, 1979; Waite, 1985). In a study reporting percentage frequencies of coliforms and other gram-negative organisms isolated from various water sources, Grabow and Du Preez (1979) found *Aeromonas hydrophila* to be the most commonly identified species with frequencies ranging from 40–58%. While the problem of false positives can be significant for total coliform results, it is less so for *E. coli* and faecal coliform counts, as Aeromonads are generally incapable of growth at 44 °C, and are oxidase positive while coliforms and *E. coli* are oxidase negative. While there are a variety of reasons for the occurrence of false positive results, no single bacterial group can be identified as being responsible for this. Instead, false positive reactions are the result of complex interactions among various genera (Hussong *et al.*, 1981).

Table 2.6 Identification of coliforms isolated from drinking water on LES ENDO agar (Mates and Shaffer, 1989)

	No. of strains	Lauryl Tryptose Broth	Brilliant Green Broth	EC Broth	% of strains
E. coli MUG +ive	36	36	36	36	23
E. coli MUG –ive	1	1	1	1	0.5
Enterobacter spp.	6	6	6	0	4
Klebsiella spp.	9	9	9	0	6
Citrobacter spp.	85	85	85	0	53
Oxidase positive organisms	23	0	0	0	14
Total	160	137	137	37	100

False negative results

As already noted, the definitions used for coliforms are not taxonomic, but are practical working definitions used for water examination purposes (APHA, 1992). It is not unexpected therefore, that some species which taxonomically belong to the coliform group will be missed using standard detection methods. These species include both anaerogenic and non-lactose fermenting strains of coliform organisms as well as non-thermotolerant strains of E. coli. Leclerc et al. (1976) have published data which show that 20% of coliforms can be lactose negative. In addition, false negative results can be obtained from apparently non-coliform colonies which on further examination are identified as coliform type species (Waite, 1985).

The phenomenon of late or non-lactose fermenting coliforms, as discussed at the beginning of the chapter, has long been recognized by water bacteriologists. However, such recognition is not allowed for in routine coliform counts, resulting in false negative results. A study of coliform recovery by membrane filtration showed that 47–61% of colonies were anaerogenic (Waite, 1985), and similar results were also obtained by Dutka (1973). The potential for such high levels of false negative results limits the applicability of the coliform group as faecal pollution indicators (Department of the Environment, 1994a; De Zuane, 1990; WHO, 1993).

2.4   THE RELEVANCE OF COLIFORMS AND FAECAL COLIFORMS AS FAECAL INDICATORS IN TROPICAL CONDITIONS

Waterborne bacterial disease has significantly declined in the more developed countries of the world, but still remains life threatening in less developed nations. The World Bank estimated that, in 1993, diarrhoea and intestinal worm infections caused by poor water supplies and bad sanitation accounted for as much as 10% of the entire disease burden of developing countries (Jones, 1994). As most less developed countries have large, undernourished populations with inadequate medical supplies, the effect of a disease outbreak is far greater, particularly in tropical areas where diseases are far more severe and diverse. At present, standard procedures for water quality assessment are used by nations in both temperate and tropical zones, despite the fact that it is widely considered that the coliform index is extremely inadequate for detecting faecal pollution in the conditions frequently found in the tropics.

It is generally assumed, based on the criteria established for indicator systems (Bonde, 1977; Feacham et al., 1983) that faecal coliforms (as their name suggests) only originate from faecal sources. It is also assumed, based on the same criteria, that they are unable to survive extra-enterally for any length of time. Several authors have reported the frequent presence of coliforms in unpolluted tropical sites and also their ability to survive for considerable lengths of time outside the intestine, thus implying that coliforms are naturally occurring in tropical waters (Bermundez and Hazen, 1988; Carrillo, Estrado and Hazen, 1985; Rivera, Hazen and Toranzos, 1988; Santiago-Mercado and Hazen, 1987). High numbers of coliforms (1.5×10^6 cfu/100 mL) have been isolated in the presence of high numbers of background organisms (2.9×10^9 cfu/ 100 mL), indicating that faecal coliforms are able to successfully compete with other organisms in the tropical environment and thus form an integral part of the normal bacterial flora of this environment (Rivera, Hazen and Toranzos, 1988).

A large proportion of coliform species in tropical waters are thermo-tolerant, that is able to ferment lactose at 44 °C (Evison and James, 1973; Santiago-Mercado and Hazen, 1987). In addition, recommended incuba-tion temperatures of 37 °C (APHA, 1992) in areas where normal temperatures average about 30 °C, will not be as selective as they are in temperate zones (Ramteke *et al.*, 1992).

Lavoie (1983) advocates the use of a direct coliform count for tropical regions, however it has been reported that *E. coli* (the target organism of the faecal coliform test), represents a smaller proportion of faecal coliform organisms than in temperate areas (Evison and James, 1973; Lavoie, 1983; Townsend, 1992). Studies of coliforms in tropical climates found that *E. coli* comprised on average 14.5% of the total coliforms isolated (Lamka, Le Chevallier and Seidler, 1980). Thermotolerant *E. coli* formed between 10 and 75% of thermotolerant coliforms (Lamka, Le Chevallier and Seidler, 1980; Lavoie, 1983). In more temperate regions, *E. coli* represents more than 90% of thermotolerant coliforms (Ramteke *et al.*, 1992). Also, the incubation temperatures required for isolation of faecal coliforms (44 °C), can present difficulties in areas with inadequate equipment and uncertain power supplies. From these observations it would appear that there is no benefit in using faecal coliforms as opposed to total coliforms as both groups give equally inaccurate results. It has been suggested that guidelines of acceptability for potable water supplies in developing countries should be set at lower levels than those generally recommended so as to encourage an incremental improvement in water quality. However, if there is no relationship between pathogen numbers and indicator organisms, there is no justifi-cation that a sample containing fewer faecal coliforms than another is any safer in terms of pathogen levels (Wright, R.C., 1982).

In view of these inadequacies, there are considerable doubts about the validity of using coliforms as indicator organisms in tropical countries. Subsequently, more suitable indicator systems are being sought (section 5.9). To date a number of possibilities have been suggested:

- **Alternative indicator systems:** the anaerobe *Bifidobacterium* has certain features which give it considerable potential as a suitable alternative. The potential of this organism is discussed in detail in section 5.5.
- **Alternative techniques:** Ramteke *et al.* (1992) have suggested the use of a P–A test for *Streptococcus*, while Manja, Maurya and Rao (1992) have developed a simple test using H_2S strips. The low cost and simplicity of this particular technique makes it an excellent candidate for use in tropical countries and it has already been implemented in certain areas (Toranzos, 1991).
- **Direct enumeration of pathogens:** the development of gene probes to directly enumerate pathogens has also been suggested (Bermundez and Hazen, 1988; Santiago-Mercado and Hazen, 1987). Although this approach is perhaps the most ideal, given the disadvantages of using indicators in tropical environments, the increased sensitivity of newer techniques also involves higher complexity and costs. Such technology is simply likely to be too expensive for developing nations which lack the capital, technology and properly trained personnel to adequately carry out such methods.

2.5   THE USE OF FAECAL COLIFORMS AS OPPOSED TO TOTAL COLIFORMS

A widely accepted approach to resolving the problems associated with using coliforms as faecal indicators, incorporated to a certain extent in the new EU Drinking Water Directive (EU, 1995), is to replace the use of total coliforms with faecal coliforms or more specifically with *E. coli* (Waite, 1985; Bonde, 1977; Dutka, 1973). It is generally accepted that the coliform index arose because early bacteriologists were unable to isolate *E. coli*, which is specifically faecal, from other coliforms. Also, according to Waite (1985), by the time the significance of *E. coli* was fully appreciated 'water bacteriologists had relied for so long on the coliform group that *E. coli* was not allowed to replace it' but was used instead to supplement it. In addition to those factors already discussed in this chapter, the argument in favour of using *E. coli* is based on the fact that the coliform group is a diverse one of different origins. While *Citrobacter*, *Klebsiella* and *Enterobacter* are all found in faeces, they are also found in extra-enteral environments, such as soil and water (Holt *et al.*, 1993). As coliform determinations are made based on a few simple morphological and biochemical tests, there is no discrimination made between coliforms of faecal and non-faecal origins.

Of the family Enterobacteriaceae, *E. coli* is the only biotype which is exclusively faecal in origin (Bonde, 1977; WHO, 1993), representing up to 95% of the Enterobacteriaceae found in faeces (Waite, 1985). It can also be easily distinguished from other faecal coliforms by the absence of urease and presence of ß-glucuronidase. The use of faecal coliforms in preference has certain advantages, in particular the reduction of high false positive results. For the assessment of drinking water quality, Bonde (1977) recommends the determination of *E. coli* numbers, since the presence of this organism certainly indicates faecal pollution. However, there are those who disagree with this recommendation (APHA, 1992; De Zuane, 1990; Wolf, 1972), pointing out that a properly treated and maintained water supply should contain no coliforms at all making it sufficient to test for total coliforms only. Even Waite (1985), who argues very strongly against the use of total coliforms, feels that for drinking waters *E. coli* testing is unnecessary. Also the numbers of faecal coliforms is generally much lower than total coliforms and therefore this test can be considered as being far less sensitive (De Zuane, 1990). In addition, *E. coli* has a low survival rate once excreted into the aquatic environment, generally surviving less than 24 hours (Borrego *et al.*, 1990; Atlas and Bartha, 1992). As *Salmonella* spp. have a lower inactivation rate than faecal coliforms there exists the risk that *Salmonella* spp. will be present long after faecal coliforms have disappeared (Borrego *et al.*, 1990). The short period of faecal coliform survival may be attributed to a number of factors, mainly:

- The physico-chemical processes of the aquatic environment such as salinity, heavy metals, sediment absorption, nutrient deficiencies, anoxic conditions, coagulation and flocculation (Atlas and Bartha, 1992; Chamberlain and Mitchell, 1978).
- The effect of certain biological, thermolabile toxins such as antibiotics that are produced by marine bacteria such as *Bacillus* and *Micrococcus* spp. These substances have a bactericidal or bacteriostatic effect on non-marine bacteria such as coliforms (Rosenfeld and Zobell, 1947).
- The effect of sunlight. In the mid-1960s solar radiation was proposed as an important factor influencing coliform decay rates. Extensive studies by Chamberlain and Mitchell (1978) later showed that coliform decay in seawater is principally the result of light induced cell damage which may in turn make them more susceptible to other factors such as predators. There is considerable variability in the sensitivity of various organisms to light intensity, coliforms being the most sensitive and which therefore tend to die off more easily.
- The action of predators and inter-species competition. Coliforms are

poor competitors at the low substrate concentrations which prevail in natural waters and so tend to be easily eliminated by competition and predation (Borrego *et al.*, 1990). Also, as with total coliforms, their usefulness as indicators of protozoan or viral contamination is limited, as is their use as absolute indicators of faecal pollution in tropical environments (Toranzos, 1991).

2.6 CONCLUSIONS

The original concept of using coliforms and faecal coliforms as indicator organisms revolutionized the whole approach to public health microbiology. The effectiveness of the coliform index has contributed to the considerable decline in the number of reported waterborne outbreaks of classical communicable disease such as cholera and typhoid fever.

While numerous studies have shown that the presence of thermotolerant coliforms generally indicates that faecal contamination has occurred, their presence in water does not always imply a health hazard. Also, the absence of coliforms does not necessarily indicate that the water is safe for consumption, in particular from enteric viruses and pathogenic protozoans. In the light of information now available on these organisms, it would appear that classic indicators such as the coliforms and faecal coliforms are increasingly inadequate to indicate their presence. It is now widely believed that the use of *E. coli* may be more appropriate for routine surveillance of drinking waters and that viruses and protozoan pathogens must be analysed separately.

One of the essential criteria for faecal coliforms and *E. coli* as indicator organisms is that they are only found in the faeces of warm-blooded animals. While this is largely true, it has been shown that they are also capable of survival in extra-enteral ecological niches, particularly in tropical zones. This is very significant given that the prevalence of bacterial diseases such as typhoid and cholera still remains very high in these regions. A proposed definition of coliforms, which is not method related, is the possession of the ß-galactosidase gene which is responsible for the cleavage of lactose into glucose and galactose by the enzyme ß-galactosidase. The ability of all Enterobacteriaceae to ferment lactose is a fundamental characteristic, and so the new simple definition will facilitate the development and acceptance of new methods of identification.

Evaluation of standard methods for the enumeration of coliforms from drinking waters

3

Despite the problems associated with the use of coliforms as indicator organisms, they remain the most commonly used bacterial indicator. There are essentially two standard methods for the enumeration of coliforms from drinking water. The multiple tube fermentation technique provides a most probable number (MPN) of coliforms in a sample, while the membrane filtration (MF) technique gives a direct colony count. Both these methods have worldwide acceptance and a long tradition of use. Prior to 1989, they were the only two methods certified by the US EPA for the detection of coliforms in water (Atlas and Bartha, 1992). As discussed in Chapter 2, the detection of coliforms has been traditionally based on the production of acid and gas in lactose based media. Both the MPN and MF techniques rely on these characteristics. This chapter outlines each technique in detail and discusses their respective advantages and limitations for coliform analysis.

3.1 THE MULTIPLE TUBE METHOD (MOST PROBABLE NUMBER METHOD)

This method was first described by McCrady in 1915 and is based on the principle of dilution to extinction (Mara, 1974). Measured volumes of one or more dilutions are added to a series of tubes containing a suitable liquid differential medium. The method is based on the assumption that each tube receives one or more test organisms in the innoculum, which will grow and produce a characteristic change associated with that organism in the medium used. The resulting pattern of positive and negative results in the dilution series is then used to obtain a statistical estimate called the most probable number.

(MPN), which is calculated by reference to probability tables devised by McCrady (1915). The MPN is expressed as the number of cells per 100 mL of sample. Detailed descriptions of all methods for coliform enumeration are given in both Standard Methods (APHA, 1992) and Report 71 (Department of the Environment, Department of Health and Social Security, and Public Health Laboratory Service, 1983) which was revised in 1994 (Department of the Environment, 1994a).

ENUMERATION OF COLIFORMS

A two stage technique is used (Figure 3.1). The first stage is known as the presumptive stage and is an estimation of the number of total coliforms in a sample. The second stage is known as the confirmatory stage.

The presumptive stage is based on the principle that coliforms ferment lactose within 48 hours at 37 °C with the production of acid and gas. A lactose based medium is used with a pH indicator to detect acid production and an inverted Durham tube to detect gas production. British practice recommends the use of Minerals Modified Glutamate Medium (MMGM) (Department of the Environment, Department of Health and Social Security, and Public Health Laboratory Service, 1983). In a comparative trial this medium was compared with Lauryl Tryptose Lactose Broth (LTLB), widely used in the USA and the medium recommended in Standard Methods (APHA, 1992) for MPN estimates (PHLS/SCA,1980a,b). The study showed MMGM to be slightly superior to LTLB, especially for chlorinated waters and for samples with low coliform counts. However, LTLB is recommended as a suitable alternative especially when rapid results are required. Both media have superseded the use of MacConkey broth. This medium is made from peptone and bile salts, both of which may vary from batch to batch in their inhibitory and nutrient properties (Mara, 1974).

It is recommended to select volumes such that the expected number of organisms will lie between 1 and 2 (Cochran, 1950). Therefore for waters expected to be of reasonably good quality, 1×50 mL and 5×10 mL volumes of sample should be used and added to equal volumes of double strength selective media. For waters of suspect or unknown quality 1×50 mL, 5×10 mL and 5×1 mL volumes of sample are used. The 1 mL volumes should be added to 5 mL of single-strength medium. For more polluted waters, the $50 \times$ mL volume should be omitted and instead replaced with 5×0.1 mL (5×1 of a 1/10 dilution). Finally for heavily polluted waters, several dilutions may be required (that is 1/100 or 1/1000) so as to maximize the chance of having some positive and some negative reactions in the dilution series (Department of the Environment, Department of Health and Social

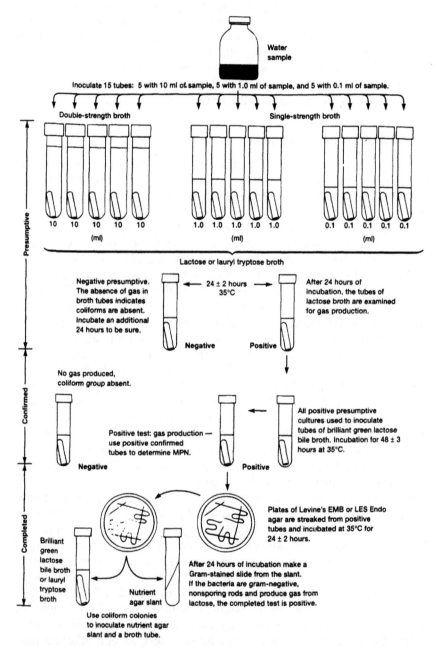

Water sample

Inoculate 15 tubes: 5 with 10 ml of sample, 5 with 1.0 ml of sample, and 5 with 0.1 ml of sample.

Double-strength broth

Single-strength broth

10 10 10 10 10 (ml)

1.0 1.0 1.0 1.0 1.0 (ml)

0.1 0.1 0.1 0.1 0.1 (ml)

Presumptive

Confirmed

Completed

Lactose or lauryl tryptose broth

Negative presumptive. The absence of gas in broth tubes indicates coliforms are absent. Incubate an additional 24 hours to be sure.

← 24 ± 2 hours 35°C →

After 24 hours of incubation, the tubes of lactose broth are examined for gas production.

Negative   Positive

No gas produced, coliform group absent.

All positive presumptive cultures used to inoculate tubes of brilliant green lactose bile broth. Incubation for 48 ± 3 hours at 35°C.

Positive test: gas production — use positive confirmed tubes to determine MPN.

Negative   Positive

Plates of Levine's EMB or LES Endo agar are streaked from positive tubes and incubated at 35°C for 24 ± 2 hours.

Brilliant green lactose bile broth or lauryl tryptose broth

Nutrient agar slant

After 24 hours of incubation make a Gram-stained slide from the slant. If the bacteria are gram-negative, nonsporing rods and produce gas from lactose, the completed test is positive.

Use coliform colonies to inoculate nutrient agar slant and a broth tube.

Figure 3.1 Summary of the multiple tube fermentation test (Prescott, Harley and Klein, 1993).

Table 3.1 Suggested sample volumes for most probable number coliform test (APHA, 1992)

Water type		Volumes (mL)					
	50	10	1	0.1	0.01	0.001
Good quality	1 tube†	5 tubes†				
Suspect or unknown quality	1 tube†	5 tubes†	5 tubes*			
Polluted waters		5 tubes*	5 tubes*	5 tubes*		
Heavily polluted waters			5 tubes*	5 tubes*	5 tubes*	5 tubes*

*Single strength.
†Double strength.

Security, and Public Health Laboratory Service, 1983); this is summarized in Table 3.1.

The procedure for the enumeration of total coliforms is:

1. After preparation of samples, suitable dilutions (as described above) are made.
2. Using aseptic technique, tubes are inoculated and incubated at 37 °C for 24 hours (Department of the Environment, Department of Health and Social Security, and Public Health Laboratory Service, 1983) or at 35 °C if following Standard Methods (APHA, 1992).
3. Tubes are examined after 24 and 48 hours. The formation of acid and gas in any tube constitutes a positive presumptive result. Any positive tubes after 24 hours should be sub-cultured to a confirmatory medium. The remaining tubes are reincubated for a further 24 hours.
4. At the end of the 48-hour incubation period all positive tubes should be recorded. The absence of acidic growth or gas formation at the end of the 48-hour incubation period constitutes a negative result. By reference to MPN tables (Table 3.2), the presumptive most probable number of total coliforms per 100 mL can be recorded.

For example, in a 15-tube test with 5×10 mL, 5×1 mL and 5×0.1 mL volumes, the number of tubes showing a positive acid and gas reaction is 4:3:1. From the tables, the presumptive MPN is 33 coliforms per 100 mL.

THE CONFIRMATORY STAGE

Any tubes which show a positive reaction at the presumptive stage should be sub-cultured to tubes of confirmatory media, either Brilliant Green Lactose Bile Broth (BGLBB) (APHA, 1992) or Lauryl Tryptose Lactose Broth (LTLB) (Department of the Environment, Department of Health and Social Security, and Public Health Laboratory Service, 1983).

Table 3.2 MPN values and 95% confidence limits for various combinations of positive and negative results when five 10-mL portions, five 5-mL portions and five 0.1-mL portions are used

Number of tubes giving positive portion out of			MPN index per 100 mL	95% confidence limits	
5 of 10 mL each	5 of 1 mL each	5 of 0.1 mL each		Lower	Upper
0	0	1	2	< 0.5	7
0	1	0	2	< 0.5	7
0	2	0	4	< 0.5	11
1	0	0	2	< 0.5	7
1	0	1	4	< 0.5	11
1	1	0	4	< 0.5	11
1	1	1	6	< 0.5	15
2	2	0	6	< 0.5	15
2	0	0	5	< 0.5	13
2	0	1	7	1	17
2	1	0	7	1	17
2	1	1	9	2	21
2	2	0	9	2	21
2	3	0	12	3	28
3	0	0	8	1	19
3	0	1	11	2	25
3	1	0	11	2	25
3	1	1	14	4	34
3	2	0	14	4	34
3	2	1	17	5	46
3	3	0	17	5	46
4	0	0	13	3	31
4	0	1	17	5	46
4	1	0	17	5	46
4	1	1	21	2	63
4	1	2	26	9	78
4	2	0	22	7	67
4	2	1	26	9	78
4	3	0	27	9	80
4	3	1	33	11	93
4	4	0	34	12	93
5	0	0	23	7	70
5	0	1	31	11	89
5	0	2	43	15	114
5	1	0	33	11	93
5	1	1	46	16	120
5	1	2	63	21	150
5	2	0	49	17	130
5	2	1	70	23	170
5	2	2	94	28	220
5	3	0	79	25	190
5	3	1	109	31	250

Table 3.2 (continued)

Number of tubes giving positive portion out of			MPN index per 100 mL	95% confidence limits	
5 of 10 mL each	5 of 1 mL each	5 of 0.1 mL each		Lower	Upper
5	3	2	141	37	340
5	3	3	175	44	500
5	4	0	130	35	300
5	4	1	172	43	490
5	4	2	221	57	700
5	4	3	278	90	850
5	4	4	345	120	1000
5	5	0	240	68	750
5	5	1	348	120	1000
5	5	2	542	180	1400
5	5	3	918	300	3200
5	5	4	1609	640	5600

Tubes are incubated at 35 °C for 48 hours (Department of the Environment, Department of Health and Social Security, and Public Health Laboratory Service, 1983) or at 37 °C for the same period (APHA, 1992). Any gas production at the end of the incubation period reads as a positive result. This confirms the presence of coliforms in the original sample. The MPN can then be determined by reading the sequence of positive tubes from the tables, as above.

COMPLETED TESTS

These tests are used on positive confirmed tubes to establish the definite presence of coliforms. US Standard Methods (APHA, 1992) recommends that confirmed tests should be used on at least 10% of positive confirmed tubes. For routine water analysis it is generally sufficient to only proceed as far as the confirmatory phase for total coliforms.

Possible tests

- Double confirmation into BGLBB for total coliforms and EC broth for faecal coliforms may be used.
- Using aseptic technique, streak a LES-ENDO agar plate or MacConkey agar plate with an inoculum from each positive BGLBB tube. All plates are incubated at 37 °C for 24 hours (APHA, 1992). On LES-ENDO typical colonies will be pink to dark red with a green metallic surface sheen. Atypical colonies will be pink, red, white or

colourless without sheen. On MacConkey agar typical colonies are red and may be surrounded by an opaque zone of precipitated bile. From each plate pick a typical (or atypical) colony and inoculate into a single-strength tube of LTLB. Tubes are incubated at 35 °C and examined after 24 and 48 hours for gas production. Gas production constitutes a positive result.

• From each MacConkey and LES-ENDO agar plate pick typical (or atypical colonies) and transfer onto nutrient agar slants. Growth on the slant is examined using the Gram stain. Gram-negative cells stain red using this stain (APHA, 1992). A summary of these tests and the positive results for total coliforms is given in Table 3.3.

Table 3.3 Summary of standard tests used for detecting the presence of coliforms in water and wastewater samples

Test name	Media	Positive test results
Presumptive	Lauryl-Tryptose Broth	Gas collected in inverted Durham tube at 35 °C by 48 hours
Confirmed	Brilliant Green Lactose Bile Broth	Gas collected in inverted Durham tube at 35 °C incubation by 48 hours
Completed	Eosine Methylene–Blue Agar	Completely dark colonies (often with metallic sheen) at 35 °C incubation by 24 hours
	Lauryl Tryptose Broth	Gas collected in inverted Durham tube at 35 °C by 48 hours
	Nutrient Agar Slant	Cells are Gram-negative non-spore-forming rods

ENUMERATION OF FAECAL COLIFORMS AND *E. COLI* BY THE MPN METHOD

The use of elevated temperatures can be used to differentiate between organisms of the coliform group from faecal sources and those from non-faecal sources. All positive presumptive tubes are inoculated into separate tubes of either BGLBB or LTLB and incubated at 44 °C for 24 hours. The production of gas confirms the presence of thermotolerant organisms. Gas formation at elevated temperatures is characteristic of faecal organisms such as faecal *Salmonella*, *Shigella* and *E. coli* strains; however, non-faecal enterobacteria such as *Enterobacter aerogenes* are also capable of growth at these temperatures.

A number of techniques can be used to differentiate faecal coliforms from other coliforms:

• **The Eijkman test:** suitable dilutions are incubated in lactose broth at 44.5 °C. Gas formation at 44.5 °C constitutes a positive result for the presence of faecal coliforms.

- **Rapid detection of** *E. coli:* this method uses two properties character-istic of *E. coli* – the ability to produce gas from lactose or mannitol at 44 °C and the ability to produce indole at 44 °C. In addition to inoculating a confirmatory medium with positive presumptive tubes a tube of tryptone water should also be inoculated and all tubes then incubated at 44 °C for 24 hours. Indole formation and gas production confirm the presence of *E. coli* in the sample (Department of the Environment, Department of Health and Social Security, and Public Health Laboratory Service, 1983). Lauryl Tryptose Manitol Broth with added tryptophan can be a suitable alternative single-tube medium (Department of the Environment, 1994a).
- **The IMViC tests:** these are tests for indole formation, the Methyl Red and Voges Proskauer reactions, citrate utilization and the production of gas from lactose at 44 °C. This battery of tests can be used to differentiate *E. coli* from non-faecal *Enterobacter* as each organism has its own characteristic pattern of positive and negative results (Table 3.4).

Standard Methods recommends the use of a faecal coliform test using specific faecal coliform agar, either EC or A1 Medium (APHA, 1992).

Table 3.4 IMViC reactions for differentiation between *E. coli* and *Enterobacter Aerogenes*

	Indole Test	Methyl Red Test	Voges Proskauer Test	Citrate Test
Enterobacter aerogenes	+	+	–	–
E. coli	–	–	+	+

STATISTICAL CONSIDERATIONS OF THE MPN METHOD

The MPN is based on certain probability formulas devised by McCrady (1915) and is an estimate of the mean density of coliforms in a sample rather than a direct count. The method has a large sampling error with the upper limit of organisms likely to be three times the MPN value and the lower limit between one third and one quarter of it. Therefore for a given estimation it is possible that the true figure will lie between these limits but this will actually only occur in approximately 5% of all such estimations (Department of the Environment, Department of Health and Social Security, and Public Health Laboratory Service, 1983). In the past confidence intervals were often published along with MPN tables. These intervals represent the uncertainty with the bacterial count for that water sample, as estimated using the MPN approach rather than uncertainty associated with the water source. This has often given rise to confusion. Hence the revised Report 71 (Department of the Environment, 1994a) has advised against their use. The MPN can now be estimated using simple computer programmes. These results are more precise, and have highlighted two issues associated with the MPN. The first has already been mentioned, that is the imprecision of confidence limits (Tillett and Coleman, 1985); and the second is the imprecision of counts when using the MPN for moderate or high bacterial densities. There is a.most probable range (MPR) of counts, all of which are equally likely.

The procedure is based on two basic assumptions:

1. The organisms are randomly dispersed throughout the sample (Poisson distribution) with no tendency of the organisms to attract or repel each other (Cochran, 1950). For this reason the importance of thoroughly mixing the sample is critical (APHA, 1992). El-Shaarawi and Pipes (1983) considered that one of the main weaknesses of the MPN procedure occurs at this fundamental stage in that most technicians and laboratory workers do not mix their samples and dilutions well enough to justify the assumption of a Poisson distribution. Also, they observed bacteria to clump together leading to errors in the final result.

2. The second assumption is that each sub-sample from the original sample when incubated into the culture medium exhibits growth whenever the sample contains one or more organisms. However, if the medium is poor, there are growth inhibiting factors present or more than one organism is required for growth, then the MPN will give an underestimation of the true density of coliforms present (Cochran, 1950). Mara (1974) noted that it is also possible to take a sub-sample which contains no coliforms, giving inaccurate results.

It is generally accepted that the MPN procedure is a method of low precision. This is to be expected from any method which does not use direct counts. The precision of any single test depends on the number of tubes used at each dilution (APHA, 1992). Unless a large number of sample portions are examined, then the precision of the MPN is rather low (Cochran, 1950; Belieff and Mary, 1993). Caution should be exercised when interpreting the sanitary significance of MPN results when only a small number of tubes are used. It is recommended that when a small tube set (three or five tubes) is used the accuracy should always be checked (Woodward, 1957). Most laboratories tend to use three or five tube sets.

SHORTCOMINGS OF THE MPN METHOD

Several investigators have reported the failure of the MPN technique to detect coliforms in drinking and other waters (Evans *et al.*, 1981b; Geldreich *et al.*, 1972 and Seidler *et al.*, 1981). A number of reasons have been suggested for this failure including interference by bacteria which are antagonistic to coliforms. Geldreich *et al.* (1972) showed that when the standard plate count exceeds 500 cells per 100 mL there is a subsequent decrease in the percentage of coliform contaminated samples. He speculated that competition among non-coliforms for lactose and the presence of bacteria antagonistic to coliforms may result in *E. coli* interference. This results in substantial numbers of coliforms being missed by the standard MPN technique (Seidler *et al.*, 1981). A modified MPN procedure was developed to document the magnitude of interference with faecal coliform detection in the standard MPN technique (Evans *et al.*, 1981b). Coliform suppression was found at all levels of the test particularly at the presumptive and confirmatory stages. The detection of coliforms in the presumptive test is dependent on the fermentation of lactose and the production of gas in sufficient quantities so as to be visible in the Durham tube. Cultural conditions which affect the activity or quantity of those enzymes responsible for gas production will ultimately affect the outcome of the presumptive test.

Another possible reason for coliform masking at the presumptive stage is coliform injury in the aquatic environment (Evans *et al.*, 1981b). Other bacteria are not as liable to stress as coliforms; therefore when a sample is inoculated into lactose-containing broth those non-coliforms present may overgrow coliforms and produce conditions unsuitable for gas production. McFeters, Cameron and Le Chevallier (1982) found LTB to have the lowest recovery of injured coliforms (average 56%). This medium is recommended in both Standard Methods (APHA, 1992) and Report 71 (Department of the Environment, Department of Health and Social Security, and Public Health Laboratory Service, 1983). Because of the problem of coliform masking it is recommended that all gas-negative, turbid, presumptive tubes be processed.

Coliform suppression at the confirmatory stage can also be a cause for the underestimation of coliforms in water samples. The reasons for failure at this stage can be attributed to the failure of some coliforms to produce gas in BGLBB and/or the failure of some coliforms to produce typical colonies on EMB agar (for completed test). The inhibitory nature of BGLBB has been well documented (Evans *et al.*, 1981b), particularly with chlorinated effluents. McFeters, Cameron and Le Chevallier (1982) observed that media containing bile salts were extremely inhibitory to injured cells and therefore had a low recovery of coliforms. Given that it is estimated that up to 90% of coliforms can become physiologically injured when exposed to aquatic environments (Bissonette *et al.*, 1975, 1977), the use of this media is questionable. Other workers have suggested that time release capsules added to MPN tubes would allow for repair of injured cells before the selective agent in the media used reached inhibitory levels (Lantz and Hartman, 1976).

A study by Evans *et al.* (1981a), using a modified MPN technique showed that the standard MPN repeatedly failed to detect the species *Enterobacter* and *Citrobacter*. These coliform genera are present at 10^5–10^6 cells per gram of human faeces. Their presence is at least a signal of inadequate water or sewage treatment or of contamination within the distribution system. The MPN is based on the ability to produce acid and gas from lactose. As discussed in Chapter 2, the production of gas from lactose is a particularly variable characteristic and not one that can be used to definitively identify coliforms. The failure of the standard MPN to detect these organisms can lead to waters being declared potable, whereas in reality they may present a sizeable public health hazard.

3.2   THE MEMBRANE FILTRATION METHOD

Until the 1950s, practical water bacteriology relied almost exclusively on the enumeration of coliforms and *E. coli* by estimating the MPN using the statistical approach originally suggested by McCrady in 1915. During the 1950s various workers developed membrane filtration into a practicable alternative to the MPN approach (Waite, 1984). The first description of molecular filter membranes for water quality control was by Thomas, Woodward and Kable in 1956, when the technique was first introduced for the routine counting of the coliaerogenes group in stored waters.

GENERAL PRINCIPLE

A measured volume of water is filtered through a membrane composed of cellulose esters. The pore size of the membrane is such that the micro-organisms are retained on or near the surface of the membrane. The membrane is then aseptically transferred to either a differential solid agar base selective for the organism sought or to an absorbent pad saturated with a suitable liquid medium. On incubation at a specific temperature for a specific time, growth will occur. It is assumed that the organisms retained by the membrane will form colonies of characteristic morphology and colour depending on the medium used. These colonies can then be counted and the number of organisms per 100 mL calculated.

PREPARATION OF EQUIPMENT AND MATERIALS

Glassware

All glassware, that is sample bottles, pipettes, graduated cylinders, containers and culture dishes, should be sterilized prior to use.

Filter membranes

Only those filter membranes which have been found to have adequate quality-control testing and certification by the manufacturer should be used. The pore diameter should be sufficient to retain coliforms on its surface. Report 71 (Department of the Environment, 1994a) and Standard Methods (APHA, 1992) recommend the use of membranes with grid marks. It is important to ensure that bacterial growth is neither inhibited nor stimulated along the grid lines.

Filtration apparatus

The filtration apparatus consists of a seamless funnel fastened to a base by a locking device or held in place by a magnetic force or gravity. Prior to use, the funnel and base should be wrapped separately and autoclaved.

Absorbent pads

If absorbent pads are to be used, they should have at least the same diameter as the membrane and be approximately 1 mm thick. They should be of high quality, be uniformly absorbent and be free from toxic substances which may prove inhibitory to bacterial growth. Prior to use, absorbent pads should be sterilized.

Media

Standard Methods (APHA, 1992) recommends that due to the need for uniformity, only dehydrated media should be used. There are a wide range of media available for the purpose of membrane filtration. These will be discussed in detail later. British practice recommends the use of Sodium Lauryl Sulphate Broth (SLSB) (PHLS/SCA, 1980b) or Teepol 610 Broth, while Standard Methods (APHA, 1992) advocates the use of either LES-ENDO agar or m-ENDO medium (Dehydrated Difco M-ENDO Broth MF or equivalent may also be used). For the enumeration of faecal coliforms from water samples, the 18th edition of Standard Methods recommends the use of m-FC medium.

Choice of volumes

Sample volumes to be filtered are chosen so that the number of colonies to be counted on the membrane will be between 10 and 100. It may be necessary with some waters to use a number of different volumes to maximize the chance of the number of colonies falling within this range. For more polluted waters it may be necessary to dilute the sample. For drinking waters where little or no coliform growth is expected, a volume of at least 100 mL should be filtered (Department of the Environment, Department of Health and Social Security, and Public Health Laboratory Service, 1983) (Table 3.5). When less than 20 mL of sample (diluted or undiluted) is to be filtered, add approximately 10 mL sterile dilution water before filtration. This increase in water volume aids in uniform dispersion of the bacterial suspension over the entire filtering surface.

Table 3.5 Suggested sample volumes for membrane filter total coliform test (APHA, 1992)

Water source	Volume (X) to be filtered (mL)							
	100	50	10	1	0.1	0.01	0.001	0.0001
Drinking water	X							
Swimming pools	X							
Wells, springs	X	X	X					
Lakes, reservoirs	X	X	X					
Water supply intake			X	X	X			
Bathing beaches			X	X	X			
River water				X	X	X	X	
Chlorinated sewage				X	X	X		
Raw sewage					X	X	X	X

PROCEDURE

1. Using a sterile forceps place a sterile filter membrane, grid side up, on the porous part of the filter base.
2. Replace funnel securely on the filter base.
3. Pour or pipette a volume of sample into the funnel.
4. Filter sample under partial vacuum.
5. With filter membrane still in place, rinse the funnel walls with sterile dilution water.
6. Remove the funnel and aseptically transfer the membrane to a petri dish containing agar or a sterile pad soaked in broth. Ensure no air bubbles are trapped between the membrane and the medium. If sterile pads are being used, any excess medium from the pad should be poured off, otherwise confluent growth may occur.
7. For different volumes of the same sample, the same funnel can be used, provided the smallest volume is sampled first. For different samples the funnel should be changed.
8. The petri dishes and membranes should then be placed in an airtight

container to prevent drying out. Alternatively, a polythene bag may be used, provided it is carefully folded over, tied and sealed.

9. For examination of membranes for total coliforms the membranes are incubated at 30 °C for 4 hours and then for 14 hours at 37 °C (Department of the Environment, Department of Health and Social Security, and Public Health Laboratory Service, 1983) (22–24 hours at 35 °C by Standard Methods (APHA, 1992).

10. For examination of thermotolerant coliforms and *E. coli*, the membranes are incubated at 30 °C for 4 hours and then for 14 hours at 44 °C (Department of the Environment, Department of Health and Social Security, and Public Health Laboratory Service, 1983). If following Standard Methods procedure, the petri dishes are incubated in a waterbath at 44.5 °C for 24 hours. Incubation temperature is critical. A solid heat sink incubator or equivalent can be used as an alternative to a water bath provided it is capable of maintaining temperatures within 0.5 °C for 24 hours.

11. An enrichment step may also be included. Samples are placed on a less selective medium or incubated at a lower temperature for a short period prior to growth under the final set of more selective conditions. For example, the use of a two-hour incubation period at 35 °C on a pad soaked with Lauryl Tryptose Sulphate Broth is carried out in the LES-ENDO procedure.

EXAMINATION AND COUNTING ON MEMBRANES

Membranes should be examined within a few minutes of being removed from the incubator, under good light and with a hand lens if necessary (Table 3.6). Colours are liable to change on cooling and when left standing.

• **Total coliforms:** the physical appearance of the colonies to be counted depends on the media used. If SLSB is used then the colonies will be yellow, whereas typical coliform colonies when grown on m-ENDO medium have a pink/dark red colour with a metallic surface sheen.

Table 3.6 Examination and counting on membranes

Organism	Media used	Colony appearance
Total coliforms	SLSB	Yellow
	m-ENDO	Pink/dark red colonies with metallic surface sheen
Faecal coliforms	m-SLB	Bright yellow
(including *E. coli*)	m-ENDO	Pink/dark red with metallic sheen
	m-FC	Blue (various shades)

• **Thermotolerant coliforms and *E. coli*:** on m-LSB, *E. coli* produces characteristic, bright yellow colonies more than 1 mm in diameter. On m-ENDO, colonies have a pink/dark red colour with a metallic sheen. Typical faecal coliform colonies on m-FC agar are various stages of blue.

CONFIRMATORY TESTS

For examination of most waters by membrane filtration presumptive results are usually sufficient. However, for potable waters it is necessary to carry out confirmation tests. Typical colonies can sometimes be produced by non-coliform organisms and can therefore lead to an over-estimation of the number of coliforms present. Any positive colonies (or a sufficient representative of them) on the membrane should be subcultured to tubes of Lactose Peptone Water (LPW) containing an inverted Durham tube. Acid and gas production after incubation at 37 °C for 48 hours indicates the presence of coliforms in the sample (Department of the Environment, 1994a).

For confirmation of *E. coli*, all typical colonies should be subcultured to tubes of Lactose Peptone Water and tubes of Tryptone Water, containing a Durham tube if information regarding gas production is required or considered relevant. The tubes are then incubated at 44 °C for 48 hours. Acid and gas production in Lactose Peptone Water and a positive Indole test gives a positive result for *E. coli*. BGLBB can be used as an alternative to Lactose Peptone Water. Few non-faecal coliforms will be observed on m-FC agar because of its selective action (APHA, 1992).

Alternative confirmatory tests include carrying out the IMViC tests (section 3.4), using a rapid multi-test identification system (section 4.3), or by using a short battery of rapid tests, that is the tests for cytochrome oxidase (CO) and ß-galactosidase activity (ONPG) respectively. Coliform reactions are CO negative and ONPG positive within 4 hours' incubation of tube culture or spot test procedure (APHA, 1992; Department of the Environment, Department of Health and Social Security, and Public Health Laboratory Service, 1983). Tests for ß-glucuronidase may also be used (Feng and Hartman, 1982).

CALCULATION OF COLIFORM DENSITY AND STATISTICAL
CONSIDERATIONS OF THE MEMBRANE FILTRATION TECHNIQUE

The number of coliforms per 100 mL can be calculated as:

$$\text{coliforms}/100\,\text{mL} = \frac{\text{coliform colonies counted} \times 100}{\text{mL sample filtered}}$$

The percentage of verified colonies can be calculated as:

$$\% \text{ Verified colonies} = \frac{\text{no. of verified colonies} \times 100}{\text{no. of coliform colonies subjected to verification}}$$

Although the statistical reliability of the membrane filter method is greater than that of the fermentation tube method, membrane counts are not absolute values. Counts on membranes subjected to variation and replicate tests using the same sample are unlikely to give the same number of colonies. Such variation is attributed to a number of factors such as the dilution of the original sample and the variability of organism density within a body of water (Tillet and Farrington, 1991). If the organisms are randomly dispersed in the sample, then the variation will be Poisson in distribution. For a colony count (C), the 95% confidence limits for the number of organisms likely to be present in 100 mL of sample can be approximated using the following formula:

$$C + 2 \pm 2 \sqrt{C} + 1$$

For example, if 100 colonies are counted on the membrane, the actual number of colonies is likely to lie between 82 and 122. However, for counts less than 20, this formula can be inaccurate. In such cases reference should be made to the Poisson tables (Table 3.7).

Table 3.7 Poisson table for estimation of coliform numbers using the membrane filtration technique (Department of the Environment, Department of Health and Social Security, and Public Health Laboratory Service, 1983)

Membrane colony count	95% confidence limits	
	Lower	Upper
1	0	4
5	0	11
10	3	18
15	7	24
20	11	30

A new approach to calculating confidence limits has recently been outlined in the revised edition of Report 71 (Department of the Environment, 1994a). This approach is based on the principles described in detail by Tillett and Farrington (1991) and takes into account the effects of dilution. In this method the confirmed colony count (y) is calculated as:

$$y = xN/n$$

where N is the total number of typical colonies per plate, n is the number of identical colonies tested and x is the number of colonies tested (n) which are confirmed. For example if 39 colonies are observed on the plate (N) and 15 colonies are randomly selected for testing (n) of which 10 were confirmed (x) then the estimated confirmed colony counts is:

$$10 \times 39/15 = 26$$

The conditional probability that y is the true count can be calculated as:

$$p^{(x-y)} = {}^{y}C_{x} \cdot {}^{N-y}C_{n-x}/{}^{N}C_{n}$$

The 95% confidence intervals can be found by studying probabilities for all possible values of y, using observed values of x. An example is given in Report 71 (Department of the Environment, 1994a).

ADVANTAGES AND DISADVANTAGES OF MEMBRANE FILTRATION

The Membrane Filtration method used for water analysis is considered to have many advantages over the Multiple Tube method. These advantages include:

- Presumptive coliform and *E. coli* counts results are available in a shorter time (18–24 hours as opposed to 48 hours).
- It is simple.
- There is considerable saving in the laboratory in the amounts of culture media, labour and glassware required.
- Larger volumes of sample may be processed.
- It is possible to carry out filtration in the field (Grabow and Du Preez, 1979; Thomas, Woodward and Kable, 1956; Windle-Taylor and Burman, 1964).
- False negative results due to development of aerobic and anaerobic spore bearing organisms are unlikely to occur.

There are also certain limitations associated with the technique, in particular:

- The method is unsuitable for turbid water as the filter become blocked too easily. It is recommended that the MPN approach should be used when turbidity exceeds five nephelametric turbidity units.
- There can be variations in the ability of individuals to recognize and identify typical coliform colonies on filters.

- Coliform growth on filters may be inhibited by high background bacterial numbers.
- Gas production is not detected. This is less critical in view of new definition of coliforms (Chapter 2).

In spite of these disadvantages, many workers consider membrane filtration (MF) to be the method of choice in the isolation of coliforms and *E. coli* from water (Dutka and Tobin, 1976; Evans *et al.*, 1981b; Grabow and Du Preez, 1979; PHLS/SCA, 1972; 1980b; Tobin, Lomax and Kushner, 1980). However, Jacobs *et al.* (1986) found MF to have the lowest recovery of coliforms compared with Standard Methods MPN and P–A techniques. The lower sensitivity of MF can be attributed to the lower survival rate of cells on a filter membrane which is possibly due to an accumulation of toxic substances on the membrane itself (Bissonette *et al.*, 1977; Jacobs *et al.*, 1986). Significantly poorer recoveries of indicator organisms by MF than by MPN have been observed, especially when examining toxic wastes and chlorinated effluents (Bissonette *et al.*, 1977; Green, Clausen and Litsky, 1977). The selective conditions of media containing inhibitory compounds and the elevated temperatures used in the case of faecal coliforms restrict the growth of chlorine-injured cells which results in a sub-maximal bacterial count (McFeters and Camper, 1983). Comparisons of faecal coliform recoveries from chlorinated effluents can show tenfold greater MPN counts (Green, Clausen and Litsky, 1977). It has been suggested that the MPN procedure provides an environment whereby injured cells can repair themselves, which would account for this difference (Stuart, McFeters and Schillinger, 1977). The problem of incomplete recovery of chlorine injured coliforms and faecal coliforms eventually reached a critical point in 1975. The dissatisfaction with MF materials and procedures led the US EPA to suggest that MF methods should be reconsidered as acceptable water quality tools in certain circumstances (McFeters and Camper, 1983).

FACTORS WHICH INFLUENCE MEMBRANE FILTRATION SENSITIVITY

Even in the early stages of membrane filtration development, it was recognized that MF had a lower recovery rate than MPN (Thomas, Woodward and Kable, 1956). While these differences can, in part, be attributed to the mathematical bias of the multiple tube method, there are a number of other factors which may contribute to the lack of sensitivity of the MF procedure, including physiological injury, the type of filter used, the type of diluent used and the type of media used.

Physiological injury

The concept of physiological injury provides some explanation for the discrepancy between membrane filtration and MPN results. Bissonette *et al.* (1975) observed that up to 90% of the total population of intestinal micro-organisms discharged into the aquatic environment may become physiologically injured as a result of exposure to new stresses. Injured bacteria have been shown to have reduced intracellular ATP, glucose transport and utilization, reduced aerobic respiration, production of secondary metabolites and resistance to disinfection, and a reduction in cell size (Singh and McFeters, 1992). Zaske, Dockins and McFeters (1980) found that even short term exposure to stress in water causes cellular envelope damage, which in turn results in a greater susceptibility to secondary stresses such as sensitivity to exposure to certain constituents in selective media. More recent studies show that several stress conditions have been known to reduce catalase activity. One consequence of this may be the accumulation of toxic hydrogen peroxide to which injured bacteria become increasingly sensitive (Calabrese and Bissonette, 1990). Stressed or injured organisms are not killed, but become very difficult to detect thereby underestimating the coliform density and faecal pathogens in aquatic environments. Sublethal injury in the aquatic environment can be related to a number of factors including time and temperature of exposure, disinfection levels, strain of organism, concentration of nutrients, presence of heavy metal ions, antagonistic standard plate count bacteria and other unidentified chemical and physical parameters (McFeters, Cameron and Le Chevallier, 1982). To recover cells which have undergone stress it is necessary that such cells have the opportunity to repair themselves before they can multiply and divide. Bissonette *et al.* (1975) observed that when an injured population of *E. coli* was exposed to a nutritionally rich non-selective broth, increasing proportions of cells were able to repair themselves to such an extent that they became insensitive to inhibitory agents in selective media. This finding resulted in the development of various resuscitation procedures and media for the improved enumeration of injured coliforms (Bissonette *et al.*, 1975, 1977; Green, Clausen and Litsky, 1977; Stuart, McFeters and Schillinger, 1977) and subsequently led to the addition of a new section describing various resuscitation techniques in the 15th edition of Standard Methods (Stressed Organism Recovery, section no. 921) (APHA, 1976).

The addition of an enrichment step has certain limitations for the MF technique in that more time, equipment and manpower are required for analysis. Evans, Seidler and Le Chevallier (1981) found that the addition of a resuscitation step to the MF technique did not increase the number of typical colonies recovered from untreated surface water or chlorinated drinking waters and therefore in their opinion served no advantage in terms of increased coliform recovery.

The type of membrane used

An often unsuspected variable in coliform and *E. coli* estimates is the type of membrane filter used. Many investigators have examined the recovery of total and faecal coliforms from different water sources using a variety of commercially available membranes (Brodsky and Schiemann, 1975; Dutka, Jackson and Bell, 1974; Green, Clausen and Litsky, 1975; Lin., 1976, 1977; Presswood and Brown, 1973; Tobin and Dutka, 1977; Tobin, Lomax and Kushner, 1980). These studies showed that there are highly significant differences between various brands of membranes in their ability to recover bacteria from various sources. Sladek *et al.* (1975) suggested that surface pore morphology may be a key factor in explaining the different results obtained using different membrane filters. He found that the greater the surface pore size, the better the recovery rate. However, neither Tobin, Lomax and Kusher (1980) nor McFeters, Cameron and Le Chevallier (1982) found any such relationship.

More recent studies by Brenner and Rankin (1990) would suggest that the differences in recoveries observed from different membrane filters may be attributed to the quality of membrane used. A total of 142 lots of membrane filters from 13 manufacturers were screened for defects. Membranes were rated acceptable if there were no significant defects, marginally acceptable if a slight defect was present and unacceptable if obvious, severe or multiple defects occurred. Results from this membrane filter lot screening study showed that no manufacturer produced defect-free filters (Table 3.8). Types of defects included partial or complete inhibition of colony development at grid lines, abnormal spreading of colonies, growth in and along the grid lines, non wetting areas, brittleness, decreased recovery and severe wrinkling. Although the number of lots tested from individual manufacturers was small in most cases and may or may not have been representative of their product, the high percentage of unacceptable lots indicates the generally poor quality of membrane filters and the need for better quality control during manufacture. These findings have serious implications for the accuracy of the membrane filtration method and may account for the differences observed between membrane filtration and other coliform enumeration methods.

The type of diluent used

Diluent composition, exposure time and temperature can greatly influence the enumeration efficiency of both stressed and non-stressed coliforms. McFeters, Cameron and Le Chevallier (1982) examined the

Table 3.8 Acceptability of 0.45 μm membrane filter lots from several manufacturers by the pure-culture screening test (Brenner and Rankin, 1990)

Manufacturer	Total no. lots tested	No. of acceptable MF lots	No. of marginal MF lots	No. of unacceptable MF lots
AMF Cuno	1	0	0	1
Amicon	4	0	0	4
Cehman Sciences	37	8	5	24
Micro Filtration Systems	5	2	0	3
Millipore Corp.	59	26	2	31
Micron Separations, Inc.	6	1	0	5
Nagle Co.	2	1	0	1
Nuclepore Corp.	8	0	2	6
Oxoid Ltd.	2	0	0	2
Schleicher and Schuell, Inc.	5	0	1	4
Sartorious	3	1	2	0
Whatman, Inc.	6	3	2	1

effects of diluents on the recovery of injured coliforms. They found that if diluents are maintained at 4 °C, their composition and exposure times are likely to have a minimal effect on enumeration efficiency; however, increased exposure times and higher temperatures can result in much lower recoveries. This study also showed that the addition of small amounts of organic materials augments the recovery of injured coliforms.

Media composition

There are a multitude of media available for membrane filtration enumeration of coliforms and faecal coliforms. Each new development or modification claims to be more selective, more able to resuscitate damaged organisms or designed to overcome specific sample conditions (Dutka and Tobin, 1976). At present m-ENDO and m-ENDO-LES are the media specified in Standard Methods (APHA, 1992) for the enumeration of coliforms by the MF procedure, whereas Report 71 recommends the use of Sodium Lauryl Sulphate Broth or Teepol 610 (Department of the Environment, 1994a).

Most media designed to select for a particular organism contain one or more inhibitory substances which suppress unwanted background flora but may also inhibit some of the organisms being selected for. If a portion of the target population is repeatedly not recovered from sample to sample, then this constant error becomes characteristic of the method, resulting in a constant under-estimation of the organisms present in the sample (Dufour, Strickland and Cabelli, 1981). This is especially a problem with injured coliforms. In a survey of over 20 media commonly used for coliform analysis, McFeters, Cameron and Le Chevallier (1982) found that the majority of media tested recovered less than 30% of injured coliforms (Table 3.9). The media are arranged in three groups based on the response of the injured cells.

Some of the media most frequently used, including m-ENDO, ranked lowest in the survey, with recoveries ranging from 5–66%. This low recovery rate is related to the concentrations of the selective agents, bile

Table 3.9 Media and the recovery of injured and healthy coliforms from water (McFeters, Cameron and Le Chevallier, 1982)

Medium	Recovery range* (%)	
	Injured	Healthy
Group I		
Triple sugar iron	181	106
Nutrient alginate	125	88
Minerals modified glutamate	99	106
Tergitol 7	86 (71–101)	99
Boric acid	84	92
TLY + 0.1% Tween 80	72	ND[‡]
Group II		
Lactose broth	72 (47–98)	102
m-ENDO	66 (30–102)	93
Lauryl tryptose	56 (34–79)	98
Levines EMB	42 (37–47)	119
3V	39	95
Purple serum	38	56
EE	38	106
Brilliant green bile 2%	34 (18–51)	106
Deoxycholate lactose	26	94
Group III		
Eosin methylene blue	24 (7–42)	102
Violet red bile	12	99
m-FC at 44.5 °C	7 (4–10)	105
MacConkey	5	97
GN	4	71
TLY-D	2	82
XLD	0	40

*(Per cent recovery) = {[(Colony Forming Units (cfu) on selective agar)/(cfu on TLY[†])] × 100}. Injury was 90–99%. The range for injured coliforms is calculated from seven repetitions using five coliform species over a one-year period.
[†]TLY agar: Tryptose Soy Broth minus glucose and supplemented with 1% lactose and 0.3% yeast extract plus 1% agar.
[‡]ND: not determined.

salts and deoxycholate, both of which are found in most available MF media and which are extremely inhibitory to injured coliforms at concentrations greater than 0.05%. The m-ENDO agar contains 0.1% deoxycholate.

In recognition of the inhibitory nature of some of the compounds used in MF media, other methods have been examined as possible substitutes (Le Chevallier, Jakonski and McFeters, 1984). Substances such as monensin (Freier and Hartman, 1987), Tergitol (Le Chevallier, Cameron and McFeters, 1983) and peroxide degrading compounds (Calabrese and Bissonette, 1988) have been suggested as potential alternatives. The most recent development in MF techniques has been the proliferation of media containing fluorogenic and chromogenic substances (Berg and Fiskdal, 1988; Brenner *et al.*, 1993; Covert *et al.*, 1989; Freier and Hartman, 1987; Lewis and Mak, 1989). The 18th edition of Standard Methods now includes a test for coliforms which uses chromogenic media (APHA, 1992). For example, a new medium m-Coli Blue 24 permits enumeration of total coliforms and *E. coli* populations in 24 hours or less. Developed by the Hach Company in Iowa, the total coliform colonies appear red and *E. coli* colonies blue, and are used on a presence and absence basis (Grant, 1996).

3.3 CONCLUSIONS

The provision of a potable water supply depends to a large extent on the ability to monitor for coliforms and faecal coliforms. In the bacteriological examination of drinking water, emphasis is often placed on frequent sampling and simpler, cheaper tests as opposed to occasional sampling by more expensive methods. As both the multiple tube methods and membrane filtration techniques are recognized as being relatively simple and inexpensive they have tended to become the mainstay of microbiological water assessment worldwide. Despite their widespread use, it is recognized that both methods suffer from inherent faults; the MPN is fundamentally inaccurate, whereas the sensitivity of the MF procedure can be affected by a number of variables. Both these methods are based on the ability of coliforms to produce acid and gas from lactose-based media. However, as 10% of coliforms do not ferment lactose, the requirement for the demonstration of production of gas is no longer the principal criterion used to distinguish coliforms from other organisms. There are also other considerations which limit the usefulness of both methods, namely:

• the time required for a health-based answer, which can take as much as 72 hours;

- the inability to differentiate coliforms from faecal coliforms without further tests;
- the subjective nature of interpreting the methods;
- stressed and injured bacteria, especially after disinfection, are very difficult to detect resulting in underestimation of both total and faecal coliform numbers – an enrichment step to allow injured coliforms to recover before isolation and enumeration is required, which significantly increases the time for analysis.

While it is accepted that no method is entirely without error, increasing pressures on diminishing water supplies and higher consumer expectations makes it important to have full confidence in the methodologies used if the absence or low levels of coliform organisms is to be interpreted in terms of safety.

There are increasing uncertainties about the validity of both the Membrane Filtration and Multiple Tube methods. Attempts are being made to overcome these problems through the development of new methods. These developments are considered in Chapters 4 and 5.

Alternative techniques for the isolation and enumeration of coliforms and *E. coli* from drinking water

4

Despite slight variations in the microbiological parameters used by different countries, similar plating procedures, most probable number methods (MPN) and membrane filtration (MF) techniques have been applied worldwide in coliform and *E. coli* detection systems (Frampton and Restaino, 1993).

However it is now widely recognized that these enumeration techniques suffer from inherent limitations, some of which have been discussed in the last chapter. This realization has stimulated research in two major directions. One is the development of other indicator systems; this will be discussed in Chapter 5. The other is the development of new techniques to estimate *E. coli* and/or faecal coliform populations (Dutka, 1979), particularly rapid methods, as it is now considered that the current time required for a completed result using standard plating procedures (72 hours) is too long for a health-based response. This chapter explores these developments and the impact they have had, if any, on changing the direction of water quality assessment.

4.1 THE PRESENCE–ABSENCE TECHNIQUE

The presence–absence (P–A) technique was developed by Clark in 1968 to provide 'a more economical device for coliform analyses' (Clark, 1968). The test is a modification of the MPN method in which a larger water sample (50 or 100 mL) is incubated in a single culture bottle containing P–A medium. Samples are incubated at 35 °C for up to 48 hours. A small inoculum is then transferred to a tube of Brilliant Green Bile Lactose Broth. Total coliform confirmation requires organisms to produce acid and gas within 48 hours at 35 °C. As only a single vessel is used, there is no information about the number of coliforms in the sample, only an indication of whether they are present or not (Clark, 1980).

Clark (1968, 1969) compared results of the P–A test with other coliform detection techniques and found it to be a more sensitive method requiring significantly less effort and expense to perform. Results were not affected by high background counts as test conditions permitted sufficient coliform growth to give a presumptive positive reaction. A comparison by Jacobs *et al.* (1986) with MF and a 10 tube MPN showed that the MPN method detected 82%, the MF method 64% and the 100 mL P–A bottle 88% of the total coliforms present in the samples. A similar study sponsored by the US EPA also found that the P–A test detected more coliform positive samples than either of the two other methods (Caldwell and Monta, 1988). These were also the findings of a further comparative study by Rice, Geldreich and Read (1989). Pipes *et al.* (1986) have published data which shows more samples positive by MF than by P–A; however, the difference was shown not to be statistically significant.

In Canada the P–A test has been used successfully since 1969 (Clark, 1980) and in England a version of the test has been used for many years for the decentralized testing of distribution waters (Furness and Holmes, 1987). In an attempt to address the shortcomings presented by the standard MPN and MF techniques, the US EPA have now decided to adopt a presence–absence concept for the detection of coliforms in water. This concept places judgement on the microbiological quality of the numbers of samples which have coliform numbers present (Federal Register, 1989) and has been discussed in more detail in Chapter 1. As a result a P–A procedure is now listed in the USA as an accepted standard method (APHA, 1992), where it is recommended that the technique is used for routine sample submissions from distribution systems or water treatment plants. In the UK an evaluation of a number of P–A tests for coliforms and *E. coli*, including Fluorocult LMX Broth (Merck), Colitrace (Bradsure Biologicals), Colilert (Palintest) and Bacto P–A Broth (Difco), has been published under the Department of the Environment series *Methods for the examination of waters and associated materials* (Lee, Lightfoot and Tillett, 1995). The study concludes that there is no P–A test that is best at all locations for both coliforms and *E. coli*, and as there can be marked ecological differences between sources it is important that particular P–A tests are validated in each geographical area before use.

4.2   DEFINED SUBSTRATE TECHNOLOGY (ENZYME DETECTION METHODS)

Chromagens and fluorogens, substrates that produce colour and fluorescence respectively upon cleavage by a specific enzyme, have been used for many years to detect and identify coliform bacteria including *E. coli* (Bascomb, 1987; Manafi, Kneifel and Bascomb, 1991). These enzymatic assays may constitute an alternative method for enumerating indicator organisms which is specific, sensitive and rapid (Bitton, 1994).

Total coliform detection is based on detection of ß-galactosidase. This enzyme catalyzes the breakdown of lactose into galactose and glucose and can be used for enumerating the coliform group within the Enterobacteriaceae family (Manafi, Kneifel and Bascomb, 1991). The assay is based on a biochemical reaction in which galactosidase cleaves the substrate o-nitro-phenyl-β-D-galactopyranoside (ONPG) to produce yellow nitrophenol which absorbs light at 420 nm (Apte *et al.*, 1995; Clark *et al.*, 1991). Other substrates for ß-galactosidase which may also be used include p-nitrophenyl-β-D-galactopyranoside (PNPG), indoxyl-β-D-glucuronide (IBDG) and 5-4-chloro-3-indoxyl-β-D-glucuronide (X-gluc) (Hofstra and Huis In't Veld, 1989).

The detection of *E. coli* is based on detection of ß-glucuronidase activity. A number of studies have shown that 90–7% of *E. coli* produce ß-D- glucuronidase (GUD) (Feng and Hartman, 1982) (Table 4.1). This enzyme is not fully specific to *E. coli*, with 40–67% of *Shigella* spp., 17–29% of *Salmonella* spp. and a few *Yersinia* spp. also showing GUD activity (Frampton and Restaino, 1993). In this assay, a substrate, generally 4-methylumbelliferyl-ß-D-glucuronide (MUG) is hydrolyzed by glucuronidase to produce a fluorescent end product (methylumbelliferone) when irradiated by long wave ultra-violet light (Clark *et al.*, 1991). Using this assay an indication of the presence of *E. coli* can be obtained after 24 hours.

A number of commercial ONPG-MUG preparations are now available for routine water analysis: Colilert (Access Analytical, Branford, CT), Coliquick (Hach Co., Loveland CO) and Colisure (Millipore Co., Bedford, MA). These can be used in either a MPN or

Table 4.1 Presence or absence of GUD activity in 198 strains of Gram-negative bacteria (Feng and Hartman, 1982)

Organism	No. tested	No. positive (%)
Citrobacter spp.	4	0 (0)
Enterobacter spp.	9	0 (0)
E. coli	110	106 (96)
Enterotoxigenic *E. coli*	10	10 (100)
Klebsiella spp.	11	0 (0)
Proteus spp.	4	0 (0)
Pseudomonas spp.	8	0 (0)
Salmonella spp.	35	6 (17)
Serratia spp.	2	0 (0)
Shigella spp.	5	2 (40)

P–A format. All ingredients are in powder form (in test tubes for the quantitative MPN method and in containers for the P–A analysis). A measured amount of water is added to each tube or container and the powder is dissolved resulting in a colourless solution. The tubes are then placed in an incubator for 24 hours at 35 °C. Test tubes with total coliforms will be yellow. Tubes are then exposed to a hand-held fluorescent light. Those tubes containing E. coli will fluoresce brightly (Edberg, Allen and Smith, 1989). The specificity of the method eliminates the need for confirmatory and completed tests (Edberg et al., 1988).

A National Evaluation of the Colilert test was sponsored by the US EPA and the American Water Works Association Research Foundation. Distribution system water from ten geographical areas representing a broad range of sources was tested. This study compared the performance of the Colilert System with that of Standard Methods MPN (quantitative) (Edberg et al., 1988) and P–A (qualitative) (Edberg, Allen and Smith, 1989) techniques. The evaluation showed that:

- The Colilert system was as sensitive as Standard Methods MPN and P–A techniques specifically enumerating 1 total coliform/100 mL.
- It simultaneously enumerated E. coli (1/100 mL) in the same analysis.
- It was not subject to false positive and false negative results by heterotrophic bacteria.
- Confirmatory tests were not required.
- Results were very easy to interpret.
- The test was very easy to inoculate.
- The test was cheaper than other commonly used methods.

However, the National evaluation also recognized certain limitations of the Colilert system. These are:

- Injured coliforms can give a weaker yellow colour making interpretation difficult. It is recommended that any yellow colour present after 24 hours be interpreted as a positive result and no colour represents a negative result (Edberg et al., 1988).
- The Colilert test has proved refractory to high densities of background heterotrophs. However, an area of growing concern is the activity of ß-galactosidase containing non-coliforms such as Aeromonas and Flavobacterium species which can yield false positive results (Covert et al., 1989; Edberg et al., 1988). Studies by Covert et al. (1989) showed that extension of the incubation time beyond a 28-hour period produced false positive ONPG Colilert tubes due to the presence of A. hydrophila and Pseudomonas spp. Consequently it is recommended that results read after 30 hours should be confirmed by a brilliant green bile lactose broth, other confirmation methods or species identification (Edberg et al., 1988).

- The national evaluation of the Colilert test was limited to drinking water distribution samples and therefore can not be applied to other waters unless the efficacy of the test has been established with that particular sample type (Covert *et al.*, 1989; Edberg *et al.*, 1988).

Various comparative studies have demonstrated that the Colilert MPN and P–A tests are comparable for the detection of total coliforms from drinking waters (Covert *et al.*, 1989; Cowburn *et al.*, 1994; Edberg, Allen and Smith, 1989; Edberg *et al.*, 1988; 1990; Gale and Broberg, 1993; Lewis and Mak, 1989; Olson *et al.*, 1991). On the basis of these observations the Colilert system was proposed by the US EPA as an alternative test procedure for the analysis of total coliforms and *E. coli* in drinking water (Federal Register, 1989). In June 1990 the US EPA added the following MUG tests to the Federal Register as proposed tests for the enumeration of *E. coli* from drinking water: (1) EC and MUG, (2) nutrient agar and MUG, and (3) the minimal medium ONPG-MUG (Colilert) test (Clark *et al.*, 1991). Tests (1) and (2) were approved in January 1991 (Pontius, 1993). However, the final incorporation of test (3) into the Federal Register was deferred, because while there is general agreement regarding the ability of enzyme detection systems to detect total coliforms in drinking waters, the same cannot be said about the detection of *E. coli*. Edberg and colleagues (Edberg, Allen and Smith, 1989; Edberg *et al.*, 1988) found Colilert to be a reliable indicator of *E. coli* based on the isolation and identification of the organism from MUG positive samples. However, further studies by Clark *et al.* (1991) found statistically significant differences between standard methods MF and Colilert in the isolation of *E. coli* from treated water samples, particularly oxidant stressed organisms. Similar observations of reduced sensitivity for *E. coli* detection in other studies have questioned the use of Colilert for *E. coli* (Gale and Broberg, 1993; Lewis and Mak, 1989).

Clark *et al.* (1991) found false negative occurrences of 12–19% for untreated waters and 61–81% for treated waters. The occurrence of false negative samples could be attributed to impaired substrate specificity and sensitivity for *E. coli* or to sub-lethal injury. The frequency of such false negatives suggests that in many instances the presence of *E. coli* could be missed and therefore the water quality will be underestimated. In addition, it has been found that up to one third of *E. coli* are not fluorogenic. The US EPA recognized this by not approving the Colilert product as a test for *E. coli* until the problem of false negatives could be corrected and it could be verified that the test could detect low levels of stressed bacteria (Pontius, 1993). More recently, studies evaluating a range of enzyme detection products for the recovery of chlorine treated *E. coli* have found them to be comparable to Standard Methods (Covert *et al.*, 1992; McCarthy, Standridge and Stasiak, 1992;

McFeters *et al.*, 1993), especially in the recovery of injured bacteria (Fricker and Fricker, 1994). The Colilert test was finally approved for use in *E. coli* testing in January 1992 (Pontius, 1993). The majority of studies in enzyme detection techniques have taken place in the USA, however, more recently there have been a number of studies in the UK (Gale and Broberg, 1989; Lee, Lightfoot and Tillett, 1995). The study by Lee, Lightfoot and Tillett (1995) found Colilert and Fluorocult LMX Broth detected more coliforms than the standard membrane filtration technique. The picture was less clear for *E. coli* with Colilert not performing as well as Fluorocult LMX or Colitrace, although performance varied with differing water sources. It was concluded that trials using a range of P–A tests should be carried out to identify the most appropriate test for any specific water source.

In addition to commercial preparations, chromogens and fluorogens can also be incorporated into a variety of media to enumerate total coliforms and *E. coli* from water. The chromogent (usually MUG) can be incorporated into the confirmatory broths for both MPN and MF procedures (Frampton and Restaino, 1993). Additionally these substances can be incorporated directly into membrane filtration media (Berg and Fiskdal, 1988; Brenner *et al.*, 1993; Covert *et al.*, 1989; Freier and Hartman, 1987; Lewis and Mak, 1989; Mates and Shaffer, 1989). Some of these allow for the simultaneous detection of total coliforms and *E. coli* (Brenner *et al.*, 1993; Grant, 1996; Petzel and Hartman, 1985). However, samples containing large numbers of competing Gram-negative microbes may overgrow and interfere with fluorescent detection of *E. coli* colonies (Frampton and Restaino, 1993). More recently a media containing chemiluminometric substrate has been described by Van Poucke and Nellis (1995). While this approach gives results in a shorter time, it is subject to considerable interference by aeromonads.

Since 1982 there has been a considerable progression in the development of defined substrate methodology. Defined substrate technology has several advantages over traditional testing protocols including simplicity, speed, specificity, ease of inoculation and interpretation. It has been shown to compare favourably with conventional techniques. However, it is not without its limitations, particularly false-positive and false-negative identifications, an area which still requires some consideration, as well as being non-quantitative.

4.3 RAPID METHODS

As has been already stated, there is a recognized need for rapid methods particularly in emergencies, when there is an urgent need to determine water quality. A summary of rapid detection methods for bacteria, not necessarily coliforms, is given in Table 4.2 although they

(a) Normal seven-digit code 5 144 572 = *E.coli*

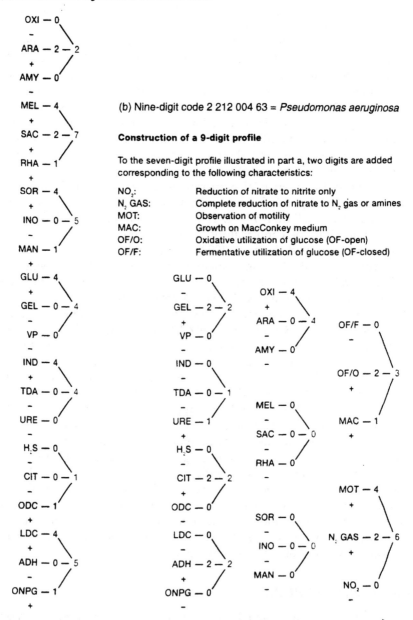

(b) Nine-digit code 2 212 004 63 = *Pseudomonas aeruginosa*

Construction of a 9-digit profile

To the seven-digit profile illustrated in part a, two digits are added corresponding to the following characteristics:

NO$_2$:	Reduction of nitrate to nitrite only
N$_2$ GAS:	Complete reduction of nitrate to N$_2$ gas or amines
MOT:	Observation of motility
MAC:	Growth on MacConkey medium
OF/O:	Oxidative utilization of glucose (OF-open)
OF/F:	Fermentative utilization of glucose (OF-closed)

Figure 4.1 The API Profile Number. The conversion of API 20E test results to the codes used in identification of unknown bacteria. The test results read top to bottom (and right to left in part b) correspond to the seven- and nine-digit

are not considered sensitive enough for potable water testing. Many are tedious and because they require specialized reagents and techniques are not suitable for routine water analysis (APHA, 1992).

A rapid seven-hour faecal coliform test was developed by Reasoner, Blannon and Geldreich, (1979) to detect faecal coliforms in water. This membrane filter test uses a lightly buffered lactose medium (m-7-hr-FC Medium) combined with a sensitive pH indicator. After incubation at 44 °C for seven hours, faecal coliform colonies appear yellow against a light purple background. These investigators found the test to be suitable for the examination of surface waters and unchlorinated sewage. A later evaluation of the method by Barnes *et al.* (1989) found it compared favourably with the Standard Methods MPN technique (APHA, 1992) when used to quantify faecal coliforms in potable and freshwater samples. The procedure was found not to be suitable for salt waters. The evaluation concluded that, although certain compromises have to be made between sensitivity and speed when enumerating coliforms by rapid methods, this method is ideal for emergency situations where rapid determination of water quality is necessary.

Numerous commercial identification systems (kits) are also available for the rapid identification of Enterobacteriaceae and other bacteria. One of the more popular systems currently in use is the API 20E (Bio Merieux UK Ltd, Basingstoke). This consists of a plastic strip with 20 microtubes containing dehydrated substrates that can detect certain biochemical characteristics: ONPG, arginine dihydrolase, lysine decarboxylase, ornithine decarboxylase, citrate utilization, H_2S production, urease production, tryptophan deamination, indole production, acetoin production, gelatin hydrolysis, fermentation of glucose, mannitol, inositol, sorbitol, rhamnose, sucrose, melibiose, amygdalin, arabinose and production of oxidase. The test substances are inoculated with bacteria in sterile physiological saline. Some tests may have to be overlaid with mineral oil to obtain the correct gaseous conditions. Results are usually available after 24 hours incubation and the 20 test results are converted to a seven- or nine-digit profile number (Figure

---

codes when read in the right to left order. The tests required for obtaining a seven-digit code take an 18–24 hour incubation and will identify most members of the Enterobacteriaceae. The longer procedure that yields a nine-digit code is required to identify many Gram-negative non-fermenting bacteria. The following tests are common to both procedures: ONPG (ß-galactosidase), ADH (arginine dihydrolase), LCD (lydine decarboxylase), ODC (ornithine decarboxylase), Cit (citrate utilization), H_2S (H_2S production), URE (urease), TDA (tryptophan deaminase), IND (indole production) VP (Voges Proskauer test), GEL (gelatin liquefaction), Fermentation of glucose (GLU), mannitol (MAN), inositol (INO), sorbitol (SOR), rhamnose (RHA), sucrose (SAC), melibiose (MEL), amygdalin (AMY), arabinose (ARA) and the oxidase test (OXI) (Prescott, Harley and Klein, 1993)

Table 4.2 Special rapid techniques (APHA, 1992)

Microbial group	Rapid method	Test time (h)	Sensitivity (cells/mL)
Non-specific microflora	Bioluminescence	1	100 000
	Chemiluminescence	1	500 000
	Impedence	3–12	100 000
	Colorimetric	0.02	10 000
	Epifluorescence/fluorometric	<1	several
Faecal coliforms	Radiometric	4–5	2–20
	Glutamate decarboxylase	10–13	0.01–500 000
	Electrochemical	1–7	1 000 000
	Impedence	6–12	200–100 000
	Gas chromatography assay	9–12	> 50
	Colorimetric	8–20	5–130 000
	Potentiometric	3.5–15	0.1–> 10 000 000
Gram-negative bacteria	Limulus assay	2	500–3000
	Fluorescent antibody	2–3	

4.1). This profile number is then used with either a computer or a book called the *API Profile Index* to find the name of the bacterium (Holmes and Costas, 1992; Prescott, Harley and Klein, 1993). A study by Holmes and Costas (1992) found that 88% of all test strains were correctly identified using the API system. They also reviewed a number of other test kits which had varying results (Table 4.3). While these kits are very convenient and relatively sensitive, cost is a considerable factor which must be taken into consideration.

There is increasing interest in the use of impedance technology for the rapid detection of coliforms and *E. coli* in water samples. The

Table 4.3 Accuracy of a number of available identification systems (Holmes and Costas, 1992)

Identification kit	% of correct identifications	% of incorrect identifications	% of organisms not identified
API (Bio Merieux UK Ltd, Basingstoke)	88	10	2
Mast ID-15 (Mast Lab. Ltd, Bootle, UK)	88	9	3
Cobas IDA (Roche Products Ltd, Welwyn Garden City)	87	7	7
Enterotube II (Roche Products Ltd, Welwyn Garden City)	90	3	6
API RAPID 20E	78	15	7
Minitek (Becton Dickinson UK Ltd, Cowley, UK)	72	16	12
Micro ID (General Diag., Morris Plains, NJ)	74	11	15

concept of electrical impedance measurement of microbial growth was originally developed by Stewart in 1899, but it was nearly 80 years later before it was used for microbial water assessment (Jay, 1991).

Impedance is the apparent resistance in an electrical circuit to the flow of alternating current corresponding to the actual electrical resistance to a direct current. When micro-organisms grow in culture media, they metabolize substrates of lower conductivity and thereby decrease the impedance of the media. For example, when the impedance of broth cultures is measured reproducible curves are produced for specific species and strains, while mixed cultures can be identified by use of specific growth inhibitors. The technique has been shown to be capable of detecting as few as 10–100 cells (Jay, 1991) and is commonly used in the food and dairy industries for a wide range of applications and tests. Recent studies have indicated that direct impedance may offer a suitable alternative to conventionally used methods for coliform detection in water samples (Colquhoun, Timms and Fricker, 1995; Irving, Stanfield and Hepburn, 1989). For example, Colquhoun *et al.* (1995) compared a direct impedance system with both the Colilert presence–absence test and the UK standard membrane filtration technique using Membrane Lauryl Sulphate Broth. It was found to correlate well with both standard methods, with little interference identified from other strains. Impedance offers a number of advantages for microbial water analysis, in particular: reduction in labour time and costs, more rapid availability of results allowing for a quicker response to water supply failures and the possibility of remote interrogation of the system to detect faecal coliforms. The main disadvantages of this method are that simultaneous detection of total coliforms and *E. coli* is not currently possible, and that samples must first be filtered so that some of the problems associated with membrane filtration may not be eliminated. The history, theory, practice and available instrumentation has been reviewed extensively by Richards *et al.* (1978) and by Fristenberg-Eden and Eden (1984).

Techniques now exist which have the potential to be used as rapid coliform tests in water analysis. Ideally such methods should be as sensitive and reliable as more conventional methods. However, at present some compromise must be made between speed and the accuracy of the result obtained. Therefore, it would seem that a considerable amount of further development is still required in order to arrive at a workable rapid method.

4.4 IMMUNODIAGNOSTIC TECHNIQUES

Immunodetection assays may also be used for the detection of coliforms and *E. coli*. Recent decades have seen a tremendous expansion in the number of such techniques available, including agglutination, Radio Immuno Assays (RIA), Enzyme Linked Immuno-Sorbent Assays (ELISA), Immuno-Fluorescent techniques (IF), Immune electron microscopy-immunogold, Immuno-Enzyme Assays (IEA), Monoclonal Antibodies (MoAbs), counter immunoeletrophoresis and neutralization (Kfir, Du Preez and Genthe, 1993). To date, there is extensive use of these tools in the water field. However, they are largely used for the identification and enumeration of pathogens and are rarely used for the routine analysis of indicator organisms (Kfir, Du Preez and Genthe, 1993). Joret *et al.* (1989) used monoclonal antibodies directed against outer membrane proteins (OMP-F protein) and those directed against alkaline phosphatase (an enzyme located in the cell periplasmic space) to detect coliforms and *E. coli*. They found anti-porin MoAbs were unable to distinguish between viable and non-viable cells. The anti-alkaline phosphate MoAbs were very specific and allowed rapid visualization. However, it was felt that sensitivity of the method needed to be determined before the method could be used for routine water analysis. Kfir, Du Preez and Genthe (1993) have reviewed in considerable depth the use of many of the immunodiagnostic tools available for water analysis. They concluded that while these methods are in many ways simpler, more rapid and less labour intensive than conventional methods, there are also considerable drawbacks associated with their use. Immunodetection methods are highly specific, but false positive results are a frequent occurrence. This is due to reactions with non-specific matter or cross-reactivity with a wide range of organisms present in a sample. Many methods do not detect viability. Because of these limitations it is unlikely that immunodetection techniques will be used on a routine basis for water analysis.

4.5 THE USE OF GENE PROBES AND PCR TO DETECT COLIFORMS IN WATER

The last two decades have seen a tremendous boom in the development of molecular methodologies which allow for the isolation of specific genes and their manipulation *in vitro*. This technology has enabled the detection of pathogenic micro-organisms in clinical and environmental samples and has therefore allowed a more precise determination of the public health risk associated with environmental samples.

GENE PROBES

Gene probes can be simply defined as small pieces of labelled nucleic acid that hybridize (become associated with DNA fragments) with a homologous complementary probe sequence (Prescott, Harley and

Klein, 1993). These pieces of nucleic acid may be labelled with a radio-active substance such as ^{32}P, and the probes detected using autoradi-ography. Alternatively, the probe may be non-radioactively labelled, which involves indirect detection of the target DNA via enzyme-linked antibodies. There is now a wide range of probes available for detection of pathogens and indicator organisms of environmental interest (Sayler and Layton, 1990). While these probes are useful, they are not sensitive enough to detect micro-organisms present in very small amounts. However, when combined with PCR their sensitivity is greatly enhanced.

PCR (POLYMERASE CHAIN REACTION)

The development of PCR in 1983 was a major breakthrough in molecular biology. The procedure has been found to be very useful in cloning, DNA sequencing, tracking genetic disorders and forensic analysis (Bitton, 1994). It has important uses in the identification of clinically important pathogens (Abbott *et al.*, 1988; Loche and Mach, 1988; Saiki *et al.*, 1988). PCR also has a number of environmental appli-cations, the most relevant to this discussion being the ability to detect indicator organisms and pathogens in water samples.

General concepts

The PCR technique was developed between 1983 and 1985 by Mullis and collaborators. The technique stimulates *in vitro* the DNA replication process that occurs *in vivo* and can result in millions of copies of the target DNA sequence being created. A single cycle of the PCR process involves three stages (Figure 4.2):

1. The target double-stranded DNA fragment is melted (denatured) by incubation at high temperatures to convert double-stranded DNA to single-stranded DNA.
2. The temperature is lowered and synthetic oligonucleotide primers, usually about 18–24 nucleotides, are added in excess. These primers anneal to the target DNA sequence.
3. Nucleotide triphosphates and a DNA polymerase are added to the reaction mixture. The primers are extended by nucleotide addition from the primers by the action of DNA polymerase (the enyzme responsible for DNA replication in cells). This enyzme (*taq* DNA polymerase) is extracted from *Thermus aquaticus*, a bacterium found in hot springs (Steffan and Atlas, 1991). Only polymerases able to function at the high temperatures employed in the technique can be used. At the end of one cycle, the targeted sequences on both

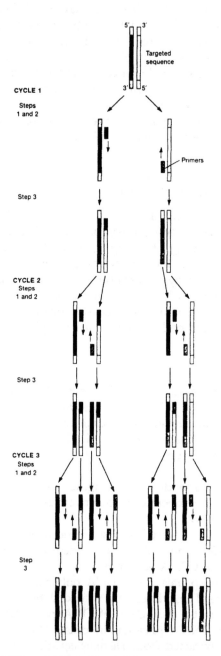

Figure 4.2 The PCR cycle (Prescott, Harley and Klein, 1993) – in three cycles the targeted sequence has been amplified to produce eight copies.

strands have been copied. Melting the product DNA duplexes and repeating the process several times results in an exponential increase in the amount of target DNA. Theoretically, 20 cycles can produce about one million copies of the target DNA sequence.

Detection of coliforms, E. coli and pathogenic organisms in water samples by PCR

A number of pathogenic organisms and indicator organisms have been successfully detected by PCR including: coliforms and E. coli (Bej et al., 1990; 1991), Shigella flexineri (Bej et al., 1990), Salmonella sp. (Bej et al., 1990; Nair et al., 1994), the Hepatitis A virus (Divizia et al., 1993), adenoviruses (Girones et al., 1993), Giardia (Mahbubani et al., 1991), Cryptosporidium (Johnson, Pienazek and Rose, 1993; Johnson et al., 1995), Klebsiella (Wong, N.A. et al., 1994) and pathogenic E. coli (Lang et al., 1994). Bej et al. (1990) were the first group to develop a genetic procedure for the recovery of coliforms and E. coli. This method involves recovery of DNA, amplification of target nucleotide sequences specifically associated with coliforms by using PCR and detection of the amplified DNA by using gene probes. The sequences chosen were a region of 'lac Z' (because conventional methods tend to be based on the detection of the lac Z gene product ß-galactosidase) and a region of the 'mal B' gene which codes for maltose transport. The 'lam B' gene was also amplified because this sequence encodes a surface protein specifically recognized by the E. coli specific bacteriophage. The rationale behind choosing these particular sequences was that PCR amplification of lac Z would occur for all coliforms, while amplification of lam B would be specific to E. coli. Initial results showed that amplification of lac Z detected E. coli and other coliforms (including Shigella) but not Salmonella and other non-coliforms, while amplification of lam B selectively detected E. coli, Salmonella and Shigella spp.

Further developments of this procedure for detecting E. coli involved amplifying regions of the uid gene which codes for ß-glucuronidase (the expression of which forms the basis of faecal coliform detection by Colilert) (Bej et al., 1991). This specifically detects E. coli and Shigella sp., including ß-glucuronidase negative strains of E. coli which generally go undetected using other detection techniques.

Multiplex PCR

The use of a multiplex PCR for amplification and detection of more than one target organism can be useful for monitoring multiple microbial pathogens in a single water sample (Bej et al., 1991). Multiplex PCR is a procedure whereby two or more primers are used to amplify two or more target sequences, and has been adapted so as to allow the detection of gene sequences associated with different groups of bacteria in environmental samples (Lang *et al.*, 1994), as it is possible that water samples will contain more than one type of microbial pathogen in addition to indicator organisms. A multiplex using *lac-Z*, *lamB*, *uid* and ss rRNA permits the simultaneous detection of a number of micro-organisms including coliforms, *E. coli* and the enteric pathogens *Salmonella*, *Shigella* and *Legionella* (Bej *et al.*, 1991).

Considerations for using PCR in the detection of indicator organisms from water samples

Current procedures for detection and enumeration of indicator organisms in water samples can take up to three days to fully process and thus potentially PCR offers a significant time saving, particularly as its increased sensitivity eliminates the requirement for additional confirmational tests (Gale and Broberg, 1993).

Sensitivity and selectivity

In order to be considered as a viable alternative to presently used methods, a PCR gene probe approach must have sufficient selectivity and sensitivity. The sensitivity requirements of any method of *E. coli* analysis in drinking water samples is one organism per 100 mL volume (Department of the Environment, Department of Health and Social Security, and Public Health Laboratory Service, 1983).

Bej *et al.* (1990) found that the sensitivity of the PCR procedure depended very much on PCR conditions but that amplification and the use of radiolabelled gene probes detected as little as 1–10 fg of genomic *E. coli* DNA and as few as 1–5 viable *E. coli* cells in 100 mL of water. Bej *et al.* (1990) also considered that PCR provided the specificity necessary for monitoring coliforms and *E. coli*. However, a more recent study by Fricker and Fricker (1994) has suggested that, while the primers used for *E. coli* detection are generally specific, those for the coliform group are not. They found that many strains did not amplify sufficiently to be detected on an agarose gel (Table 4.4). Limited experiments using enyzme labelled DNA showed that this detection system did not allow detection of all coliform strains.

Culture-requiring methods for the detection of coliforms are often limited in their ability to detect viable but non-culturable bacteria. These may include injured bacteria or bacteria in an pseudosenescent state, often associated with bacteria adapted to low nutrient conditions (Atlas and Bartha, 1992). A considerable advantage of PCR is that it may provide an answer to this problem. Bej *et al.* (1991) found that

Table 4.4 List of non-*Escherichia coli* coliform strains tested by PCR showing the number which identified as coliforms using a previously described primer set (Fricker and Fricker, 1994)

	Number tested	Coliform +ve
Citrobacter freundii	54	23
Enterobacter		
cloacae	51	43
E. agglomerans	63	54
E. amnigenus	25	6
E. intermedius	1	1
E. taylorae	1	1
Klebsiella		
terrigena	11	1
K. oxytoca	52	48
K. ozaenae	8	8
K. pneumoniae	35	31
Serratia rubideae	1	1
Leclercia adecarboxylata	4	4
Hafnia alvei	11	3*
Buttiauxella agrestis	4	4
Escherichia vulneris	1	1
Serratia odorifera	2	2*
Total	324	233

*These isolates gave two bands and would therefore be identified as *E. coli.*

PCR allowed detection of MUG-negative *E. coli* previously undetected by the Colilert test.

Increased sensitivity may also have negative aspects, in that there may be a substantial increase in false positive results (Alvarez, Hernandez-Delgado and Toranzos, 1993). PCR amplifies all DNA present as target sequence but does not differentiate between viable and non-viable cells. Subsequently there is no differentiation made between infectious and non-infectious cells and potential pathogenicity is not taken into account (Alexander and Morris, 1991). Hence a positive result does not reflect any estimate of a public health risk from a given water sample.

Quantification of PCR target sequence

Quantitative estimates of a target sequence amplified by the PCR procedure are hampered by the fact that the amount of PCR products formed increases exponentially and therefore any changes in any of the parameters which affect the efficiency of the amplification can affect the result of the reaction (Steffan and Atlas, 1991). In their review of PCR and its applications in environmental microbiology, Steffan and Atlas (1991) outline an approach which allows the PCR product to be quantified permitting estimates of indicator organisms in environmental samples.

Inhibition

PCR analysis of environmental samples requires relatively purified DNA (Steffan and Atlas, 1991). This requirement has been problematic for recovery of indicator organisms from such samples because of the presence of various interfering substances such as humic acids, metal ions and other inhibitory substances (Tsai, Plamer and Sangermano, 1993). Alexander and Morris (1991) found that if no attempt was made to clean up environmental samples, including water samples, then PCR was routinely inhibited.

Adaptability for routine use

When considering the applicability of PCR for use in routine water analysis in all types of water utility, a number of considerations must be taken into account, including the equipment and expertise required. One major limitation for the routine use of PCR has been the use of radioisotopes. Smaller water utilities lack the facilities to adequately dispose of such isotopes. However, in recent years considerable developments in this area have meant that detection methods are no longer entirely dependent on the use of radio-labelled probes. PCR non-radio-labelled detection methods can be classified into three categories:

1. the use of non-radioactively labelled gene probes;
2. the direct incorporation of label into the amplified product for use as a probe; the DNA probe is linked to enzymes such as horse radish peroxidase; alternatively, biotin or fluorescent dyes may be used;
3. the indirect incorporation of label into the PCR product for probe or capture mediated detection of amplified product (Steffan and Atlas, 1991).

The future of molecular methods in water analysis

It cannot be denied that there are many problems associated with PCR which must be addressed before this method can be routinely applied in environmental monitoring. However, such problems are to be expected as this methodology is relatively recent. The PCR procedure has enormous potential for the future of environmental monitoring. It is quite likely that with the current rate of development, its use will be in the direct detection of pathogens rather than the present indirect detection of indicator organisms.

4.6 CONCLUSIONS

The most significant limitation of standard MPN and MF methods is the time required for a definite health-based answer. Therefore any potential alternative must involve specificity, sensitivity and precision with achievement of a test result within a few hours (Geldreich, 1992). As might be expected, rapid tests that can currently be performed in less than one hour have little specificity and may include non-viable cells. Little attention has been given to these methods in this chapter, which highlights the more important developments to emerge in the last two decades. While many of these still require several hours to produce results, the time involved is still considerably shorter than that for Standard Methods and the specificity and sensitivity required is still retained.

All of the methods discussed have certain limitations, in particular false positive results and the inability to distinguish viable from non-viable cells. Of the techniques currently being developed, gene probes and PCR appear to have the most potential. This is perhaps inevitable, given the growing emphasis on molecular biology and associated techniques in the last decade. It is to be hoped that the development of multiplex probes to detect bacterial pathogens or enteroviruses will become sensitive enough to detect one organism in a litter of samples within one hour of processing (Geldreich, 1992). However, PCR is still in an early developmental stage with considerable limitations which must be overcome if this method is to become applicable on a routine basis.

Alternative indicator systems for water quality analysis

5

In addition to the development of improved techniques for the isolation and enumeration of total coliforms and faecal coliforms, realization that there are inherent problems with their use as indicator organisms has also resulted in the search for an alternative indicator system which unequivocally denotes the presence of faecal material and the existence of a potential health hazard. Over the years, several groups of organisms have been suggested as tentative alternatives to coliforms and *E. coli*. Some of these, most notably the faecal streptococci and *Clostridium perfringens*, have a history of use as long as the faecal coliforms, while the use of other groups is much more recent. This chapter examines the more important alternative indicator systems for microbial water quality analysis.

5.1 FAECAL STREPTOCOCCI

The faecal streptococci are a group of Gram-positive cocci, occurring in chains of varying length. They are both non-sporulating and non-motile and all give a positive reaction with Lancefield's group-D antisera. They grow in the presence of bile salts at 44 °C, in concentrations of sodium azide that is inhibitory to most gram-negative bacteria including coliforms (Mara, 1974). Taxonomically the faecal streptococci belong to the genera *Enterococcus* and *Streptococcus* (Devriese *et al.*, 1992; Holt *et al.*, 1993). The genus *Enterococcus* includes all streptococci that share certain biochemical properties and have a wide tolerance of adverse growth conditions (WHO, 1993). They are differentiated from other streptococci by their ability to grow in 6.5% sodium chloride, at pH 9.6 and at 10 °C and 45 °C (APHA, 1992) and include the species *Ent. avium*, *Ent. faecium*, *Ent. durans*, *Ent. faecalis* and *Ent. gallinarium*. The term enterococcus is rarely used in the field of water pollution control (Ellis, 1989). Of the genus *Streptococcus*, only *Str. bovis* and *Str. equinus* are considered to be true faecal streptococci as these are the only members of this genus to possess the group D antigen (WHO, 1993). These species are predominantly found in animal faeces (Devriese *et al.*, 1992; Geldreich and Kenner, 1969; Wheater, Mara and Oragui, 1979) while *Ent. faecalis* and *Ent. faecium* are considered to be more specific to the human gut. Other streptococcal species have also been isolated from the human gut, albeit in less numbers (Watanabe *et al.*, 1981). The group also includes other biotypes, which are ubiquitous in nature, such as *Ent. casseloflavus, Ent. faecalis* var *liquefaciens, Ent. malodoratus* and *Ent. solidarius* (Sinton, Donnison and Hastie, 1993a; WHO, 1993).

Interest in the use of faecal streptococci as a pollution indicator dates back as far as 1900, when they were found to be consistently present in the faeces of warm blooded animals and in waters associated with discharges from such animals. However, the literature showed little evidence of their application as potential indicators of human pathogens until improved methods for their enumeration and detection were developed in the 1950s (Geldreich and Kenner, 1969). The new EU Drinking Water Directive gives maximum permissible values for faecal streptococci in finished waters and bottled waters (EU, 1995). Their use as pollution indicators has been reviewed by Sinton, Donnison and Hastie (1993b).

Numerically, faecal streptococci can be just as abundant as coliforms, especially in stormwaters and specific effluents (e.g. intensive animal rearing units), but under normal circumstances they are slightly less abundant. Faecal streptococci are considered to have certain advantages over coliforms as pollution indicators:

- They rarely multiply in water (Feacham *et al.*, 1983; Geldreich, 1970; WHO, 1993).
- They are more resistant to environmental stress and chlorination than coliforms (Pipes, 1982a).
- They generally persist longer in the environment (McFeters, Bissonette and Jezeski, 1974; Vascoucelos and Swartz, 1976), with the exception of *Str. bovis* and *Str. equinus* which die off rapidly once outside the animal intestinal tract (Geldreich and Kenner, 1969). In contrast to other studies Dutka and Kwan (1980) have also observed *Ent. faecalis* to have a faster die off rate than *E. coli*. However, this observation should be treated with some caution. As Feacham *et al.* (1983) note, 'there are probably considerable inter- and intra-species variation in survival ability. Studies on mixed populations of faecal *Streptococci* and *E. coli* cannot be compared directly with studies on the survival of single laboratory maintained strains.'

The primary value of faecal streptococci in water quality examination is in situations where the coliform test is of limited value. In Britain, faecal streptococci are used to assess the significance of doubtful results from other organisms. For example, in the event of a large occurrence of coliforms in the absence of *E. coli*, the presence of faecal streptococci is used to confirm faecal contamination (Department of the Environment, Department of Health and Social Security, and Public Health Laboratory Service, 1983). The WHO (1993) recommends the use of faecal streptococci as additional indicators of treatment efficiency. As these organisms are resistant to drying, they may be valuable for routine control after laying new mains or repairs in distribution systems or for detecting pollution by surface run-off to ground or surface waters.

As *Str. bovis* and *Str. equinus* are predominantly found in warm-blooded animal faeces, this suggests that they may serve as specific indicators of non-human (warm-blooded animal) pollution (Geldreich, 1970). *Streptococcus bovis* is widely associated with cattle and other farm animals and is a relatively uncommon member of the human gut flora (Mara and Oragui, 1983). *Enterococcus faecalis* and *Ent. faecium* have both been used to indicate pollution due to faeces, although the latter is far more specific. The absence of *Ent. faecalis* should not exclude the possibility of pollution by human excreta. It is now currently felt that it is not possible to differentiate the source of faecal contamination based on speciation of faecal streptococci (APHA, 1992). However, because all the species are indicative of faecal pollution, it is not necessary to identify individual species, as a single test for the whole group is sufficient. This is particularly appropriate as many of the so-called emerging pathogens reviewed in Chapter 6 are zoonoses.

Despite their advantages as indicator organisms, there are a number of characteristics which detract from the value of faecal streptococci:

- They are less numerous than coliforms in human faeces which makes them a less sensitive indicator of human faecal contamination (Table 5.1) (Pipes, 1982a).

Table 5.1 The number of indicator bacteria commonly found in human faeces expressed as cells per gram of faeces (wet weight) (Feacham *et al.*, 1983)

Indicator	Cells/g faeces (w/w)
Bacteroides spp.	10^7–10^{11}
Bifidobacterium spp.	10^7–10^{11}
Clostridium perfringens	10^3–10^{10}
Coliforms	
Faecal	10^6–10^9
Non-faecal	10^7–10^9
Faecal streptococci	10^5–10^8

- Geldreich and Kenner (1969) found the subgroup *Ent. faecalis*, ubiquitous in nature, to be highly variable persisting longer in water than similarly exposed faecal coliforms (Figure 5.1) (Geldreich, 1970).
- There is a lack of standard methodology for their selective enumeration, with over 70 different media proposed (Pavlova, Brezenski and Litsky, 1972; Yoshpe-Purer, 1989).
- They are taxonomically and ecologically heterogeneous (Audicana, Perales and Borrego, 1995).
- Species of the group have different levels of sanitary significance (Gauci, 1991).
- Their selective enumeration is very time consuming.

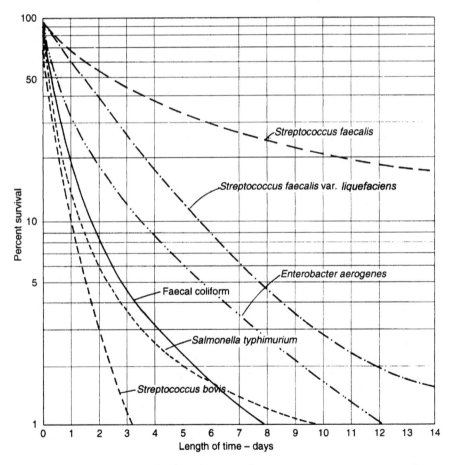

Figure 5.1 Persistence of selected enteric bacteria in storm water stored at 20 °C (Geldreich, 1970).

Further studies by Geldreich (1970) showed that *Ent. faecalis* var *liquefaciens* frequently forms a substantial portion of the low faecal streptococci densities common to good quality waters. In these waters this organism may reach significant proportions. This causes concern when interpreting faecal streptococcal numbers below 100 organisms/100 mL. Finally, as it is impossible to differentiate between faecal streptococci of faecal and non-faecal origin using standard methodologies, the use of faecal streptococci is perhaps of limited value. Poucher *et al.* (1991) have suggested that the identification of various faecal streptococci is a more viable solution as opposed to their enumeration in polluted waters.

ENUMERATION TECHNIQUES

Both the membrane filtration and most probable number techniques may be used for the isolation of faecal streptococci. As *Ent. faecalis* and closely related species are able to reduce 2,4,5-triphenyltetrazolium chloride (TTC) to formazan, a red dye, isolation methods use agar containing TTC with faecal streptococci colonies appearing red, maroon or pink due to formazan formation.

- *Membrane filtration:* the method used is similar to that used for coliforms except that the medium used is Membrane-Enterococcus Agar (MEA) with incubation at 37 °C for 48 hours for treated potable waters and for untreated water 37 °C for 4 hours followed by 44 °C for 44 hours. The more temperature-sensitive faecal streptococci can be inhibited if incubated at 45 °C even though this is more selective. In fact inhibition of temperature sensitive species may occur with a temperature rise of less than 0.5 °C. All red, maroon and pink colonies which are smooth and convex in shape are counted as presumptive faecal streptococci. Colourless colonies may be produced by some faecal streptococci. Confirmation of faecal streptococci is by sub-culture onto Bile Aesulin Agar and incubation for 18 hours at 44 °C. Faecal streptococci form discrete colonies surrounded by a brown or black halo due to aesculin hydrolysis (APHA, 1992; Department of the Environment, 1994a). This method is specified for use in the EU Drinking Water Directive (EU, 1995).
- *The most probable number technique:* the technique employed is similar to that previously described for coliforms (MPN) except that Azide Dextrose Broth (APHA, 1992) or Glucose Azide Broth (GAB) is used (Mara, 1974; Department of the Environment, 1994a). Tubes containing Azide Dextrose Broth are incubated at 35 °C and examined for turbidity after 24 and 48 hours. All turbid tubes are then subjected to the confirmation test, while tubes containing GAB are incubated at 37 °C and examined after 24 and 48 hours. Growth

and acid production is a positive result indicating the presence of presumptive faecal streptococci, although subculture into Bile Aesculin Agar is required for confirmation.

- *Other techniques:* Clark's PA technique may also be used for detection of faecal streptococci. This technique is, however, very time con-suming and awkward for faecal streptococci enumeration (Clark, 1968).

Enumeration techniques for faecal streptococci have been revised by a number of workers (Audicana, Perales and Borrego, 1995; Dionisio and Borrego, 1995). Dionisio and Borrego (1995) found membrane filtration using MEA to give the best performance characteristics of the enumeration media they evaluated in terms of recovery efficiency, precision, accuracy and specificity.

5.2   THE FAECAL COLIFORM/FAECAL STREPTOCOCCI (FC/FS) RATIO

The FC/FS ratio has been used to determine whether pollution is of animal or human origin. The basis for using the ratio can be found in the early literature. Geldreich and Kenner (1969) reported faecal coliforms to be more numerous than faecal streptococci in human faeces with a FC/FS ratio always greater than four. Conversely, faecal strepto-cocci were found to be more numerous than faecal coliforms in animal faeces with the FC/FS ratio always less than 0.7. This large difference is exploited as shown in Table 5.2.

Expected densities of faecal coliforms and faecal streptococci in human and various animal faeces, and the resultant FC/FS ratios are given in Lynch and Poole (1979) and Gray (1992). According to Geldreich and Kenner (1969), this ratio is only really valid in the first 24 hours after the bacteria have been discharged into the water. However, Poucher *et al.* (1991) have shown that even in the first few hours after collection, the FC/FS ratio is not constant, even for samples of the same origin.

Table 5.2 The FC/FS ratio

FC/FS ratio	Source of pollution
>4.0	Strong evidence that pollution is of human origin.
>2.0 <4.0	Good evidence of the predominance of human wastes in mixed pollution.
>0.7 <2.0	Good evidence of the predominance of domestic animal wastes in mixed pollution.
<0.7	Strong evidence that pollution is of animal origin.

The validity of the FC/FS ratio has been questioned considerably. McFeters, Bissonette and Jezeski (1974) pointed out that the ratios are dependent on the different die-away rates of faecal coliform and faecal streptococci. As a result of these die-away rates, the ratios will change once the faeces is excreted (Feacham *et al.*, 1983). The enterococci (*Ent. faecalis* and *Ent. faecium*) survive much longer than faecal coliforms which survive longer than either *Str. equinus* or *Str. bovis*, both of which die off rapidly when exposed to aquatic environments (McFeters, Bissonette and Jezeski, 1974). Dutka and Kwan (1980) observed that *Ent. faecalis* had a faster die-off rate than *E. coli*. This observation, if valid, would suggest that *E. coli* is more resistant to environmental stresses and therefore such results would bias the FC/FS ratio towards human faecal pollution. Wheater, Mara and Oragui (1979) observed that not all animals maintain a FC/FS ratio of less than one. Hussong *et al.* (1979) found that the FC/FS ratio seemed to be depend on diet and that it would not be sufficient to separate avian from human faecal contamination. In addition, disinfection of wastewaters appears to have a significant effect on the indicator ratio, which may lead to inaccurate conclusions. In their studies, Poucher *et al.* (1991) concluded that it is not possible to determine the source of water contamination on the basis of the FC/FS ratio.

The ratio is also almost always affected by the methods for enumerating faecal streptococci (APHA, 1992; Wheater, Mara and Oragui, 1979). So, in general, it would appear that there are too many factors which influence the FC/FS ratio for it to reliably differentiate between human and animal faecal contamination of water. In addition to those problems already mentioned, other variables such as time, temperature, pH, type of membrane used and the possibility of multiple sources of pollution may also affect the validity of the results (Department of the Environment, Department of Health and Social Security, and Public Health Laboratory Service, 1983). Consequently, the use of the FC/FS ratio is generally not recommended as a means of differentiating between different pollution sources (Poucher *et al.*, 1991; Rutkowski and Sjøgren, 1987).

5.3   *CLOSTRIDIUM PERFRINGENS*

Sulphite-reducing clostridia are anaerobic spore-forming non-motile, Gram-positive rods which are exclusively faecal in origin. Spores are very resistant and are able to withstand heating at 75 °C for 15 minutes, they can reduce sulphite to sulphide, ferment lactose and produce gas, and along with *Clostridium perfringens*, the most important member of the group, form a stormy clot in Litmus Milk Medium (Mara, 1974). They are also pathogenic, causing gas gangrene and food poisoning.

The use of *Clostridium perfringens* as an indicator organism was first proposed in 1899 by Klein and Houston (Bisson and Cabelli, 1979). Clostridial spores can survive in water much longer than either coliforms or streptococci. This persistence can indicate occasional or intermittent pollution, which then implies the need for greater frequency of sampling (Department of the Environment, Department of Health and Social Security, and Public Health Laboratory Service, 1983). Clostridial spores are also resistant to disinfection and therefore are not reduced appreciably by treatment (Table 5.3). Subsequently, they are of little value in assessing the efficiency of water treatment (Fujioka and Shizumura, 1985), however, they are of use in assessing the efficiency of filtration and the susceptibility of water resources to intermittent pollution (WHO, 1993).

Although *Clostridia perfringens* is consistently found in faecal wastes, it is not used as a faecal indicator in the USA and is only used to provide supplementary information in Europe (Bisson and Cabelli, 1980). There are a number of reasons for this. First, the spores are too resistant to chlorination to be of value in assessing drinking water quality (Pipes, 1982a; Department of the Environment, Department of Health and Social Security, and Public Health Laboratory Service, 1983). Second, they are not suitable as indicator organisms in recreational waters as sedimented spores can be resuspended by bather activity or surf action or in areas where there is significant land run-off. Spores can survive for very long periods and in such cases spores may be detected long after a pollution incidence giving rise to unnecessary, false alarms (Bisson and Cabelli, 1980; Cabelli, 1978). Last, there is the problem of finding a reliable method for enumerating *Cl. perfringens* in water. There are several methods for recovering *Cl. perfringens* from water. These are the MPN, pour plate methods and membrane filtration. All of these methods use sulphite reduction as the differential

Table 5.3 Occurrence of *Clostridium perfringens* vegetative cells and heat sensitive spores in water (Bisson and Cabelli, 1980)

Treatment plant	% recovery from wastewater			
	Pre-chlorination		Post-chlorination	
	Vegetative cells	Heat sensitive spores	Vegetative cells	Heat sensitive spores
1	6.5	50	0	14.0
2	0	12.2	0	34.7
3	13.4	–	–	–
Average	6.6	31.0	0	24.4

characteristic with the use of stormy fermentation of milk for specific identification (Mara, 1974; Bisson and Cabelli, 1980). A two-stage MPN technique is used to isolate and enumerate *Cl. perfringens*. First the water sample is heated to 75 °C for 10 minutes and after cooling is inoculated directly into Differential Reinforced Clostridial Broth and incubated at 37 °C for 48 hours. The inocula are placed in screw topped bottles partially filled with medium, which are then filled to the top with excess medium and tightly closed to ensure anaerobic conditions. The medium contains sulphite and if clostridia are present then it will be reduced and precipitated to ferrous sulphide which turns all the medium black. Confirmation of *Cl. perfringens* requires an inoculum for each positive bottle to be transferred to freshly prepared tubes of Litmus Milk which are incubated for 48 hours at 37 °C. To encourage anaerobic growth the redox potential of the medium is reduced by adding a small length of iron wire, often a small nail, that is sterilized immediately before use by heating until red hot. If *Cl. perfringens* is present a stormy clot is formed in the tube (Mara, 1974). Membrane filtration is also widely used. After preliminary heat treatment to destroy the vegetative bacteria present, a volume of the sample is filtered through a membrane that is incubated anaerobically on a sulphite-containing agar medium. Black colonies are formed by the presence of sulphite-reducing clostridia. *Clostridium perfringens* is confirmed by subculturing the colonies into a tube of Litmus Milk as described previously. There are currently a number of media available, but no method has been officially adopted by the Department of the Environment (Department of the Environment, 1994a). Fujioka and Shizumura (1985) have assessed a more practical membrane filtration method for recovery of *Cl. perfringens* from water samples using modified *Clostridium perfringens* agar (m-CP) medium. This procedure is outlined in Figure 5.2. After incubation at 45 °C on m-CP medium, typical colonies are approximately 1–3 mm in diameter, are somewhat opaque, slightly butryous in consistency and have a pale yellow colour. Upon exposure to ammonia vapours, these colonies turn a pink to red (but not purple) colour. This procedure was found to be very reliable with low numbers of false positive results. However, it is generally considered that while these methods are adequate for research and special investigations, they are time consuming and largely impractical for use on a routine basis.

5.4 BACTERIOPHAGE

Because of their constant presence in sewage, faeces and polluted waters, the use of bacteriophages (or bacterial viruses) as appropriate indicators of faecal pollution has been proposed by several authors

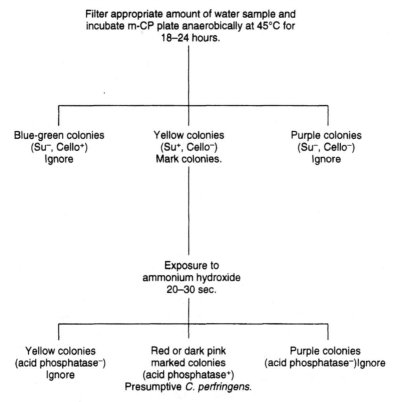

Filter appropriate amount of water sample and
incubate m-CP plate anaerobically at 45°C for
18–24 hours.

Blue-green colonies
(Su⁻, Cello⁺)
Ignore

Yellow colonies
(Su⁺, Cello⁻)
Mark colonies.

Purple colonies
(Su⁻, Cello⁻)
Ignore

Exposure to
ammonium hydroxide
20–30 sec.

Yellow colonies
(acid phosphatase⁻)
Ignore

Red or dark pink
marked colonies
(acid phosphatase⁺)
Presumptive *C. perfringens.*

Purple colonies
(acid phosphatase⁻)Ignore

Figure 5.2 Flow scheme for the m-CP procedure (Bisson and Cabelli, 1979).

(Borrego *et al.*, 1987; 1990; Dhillon *et al.*, 1976; Ratto *et al.*, 1989). These organisms have also been suggested as indicators of viral pollution. This is because their structure, morphology and size as well as their behaviour in the aquatic environment closely resemble those of enteric viruses (Geldenhuys and Pretorious, 1989; Kott *et al.*, 1974; Simkova and Cervenka, 1981; Stetler, 1984).

The use of bacteriophages as indicators of faecal pollution is based on the assumption that their presence in water samples denotes the presence of bacteria capable of supporting the replication of the phage. Two groups of phage in particular have been studied: the somatic coliphage, which infect *E. coli* host strains through cell wall receptors, and the F-specific RNA bacteriophage, which infect strains of *E. coli* and related bacteria through the F- or sex pili (WHO, 1993). A significant advantage of using bacteriophages is that they can be detected by simple and inexpensive techniques which yield results in 8–18 hours (Grabow, 1986). A proposed method for coliphage detection is outlined in the 18th edition of Standard Methods (APHA, 1992).

To date, most attention has been directed to the use of coliphages (Cornax et al., 1991; Morinigo et al., 1992). Most research has been done on the use of coliphages to mimic viruses during water treatment, and indeed coliphages and enteroviruses appeared to be removed or inactivated at similar rates during treatment processes, including chlorination (Havelaar and Hogeboom, 1984; Payment, 1991). However, their use may be limited as indicators of faecal pollution as they are not widely carried by a significant portion of the human population (Havelaar, Furuse and Hogeboom, 1986; Havelaar et al., 1990). Certain limitations associated with the use of coliphages have been reported. These include:

• Enteric viruses have been detected in their absence.
• They may replicate in natural waters under certain conditions.
• Autochthonous coliphages have been detected in unpolluted waters (Morinigo et al., 1992).

These limitations have prompted research into the suitability of other phage such as the F-specific RNA bacteriophage as indicators of faecal pollution. Studies by Havelaar et al. (1990) show that these phages occur rarely in faeces and show no direct relationship with faecal pollution. In addition, Cornax et al. (1991) observed that high concentrations of F- specific bacteriophage were found in sewage, despite low concentrations in faeces. Havelaar and Hogeboom (1988) observed similar results. They concluded that such high counts could not be explained by direct faecal impact and that some degree of multiplication must take place in the environment. While these phages cannot be recommended as indicators of faecal pollution, their presence in high numbers in waste waters and their relatively high resistance to chlorination makes them suitable as indicators of sewage pollution.

Bacteriophages of Bacteroides fragilis have also been suggested as potential indicators of human viruses in the environment (Tartera and Jofre, 1985). Bacteroides spp. are strict anaerobes and are a major component of human faeces (Table 5.1) and therefore bacteriophage active against these organisms have the potential to be suitable indicators of viral contamination.

There are a number of factors which would support this suggestion:

• Bacteriophage infected Bacteroides fragilis appear to be exclusively human in origin (Table 5.4) (Tartera and Jofre, 1987) and appear only to be present in environmental samples contaminated with human faecal pollution (Table 5.5) (Jofre et al., 1987). This may help to differentiate human from animal contamination.
• They are absent from natural habitats, which is a considerable advantage over coliphages which are found in habitats other than the human gut.

Table 5.4 Bacteriophage active against *B. fragilis* HSP 40 in faeces of various animals (Tartera and Jofre, 1987)

Animal spp.	No. tested	No. (%) of positive tests
Human	40	4 (10)
Cow	40	0 (0)
Pig	50	0 (0)
Rabbit	21	0 (0)
Mouse	28	0 (0)
Hen	20	0 (0)
Quail	10	0 (0)

Table 5.5 Presence of bacteriophage active against *B. fragilis* B40 in different environmental samples (Jofre *et al.*, 1987)

Sample type	Range of faecal coliforms/100 mL	No. of samples	% of positive samples
Water from lagoons with abundant wildlife	10^3–10^4	1	0
	10^2–10^3	6	0
	10–10^2	8	0
Surface water near urban areas	$>10^4$	50	100
	10^3–10^4	13	70
	10^2–10^3	3	66
	10–10^2	2	50
Sediments from lagoons with abundant wildlife	10^2–10^3	7	0

- They are unable to multiply in the environment (Tartera, Lucena and Jofre, 1989).
- Their decay rate is very similar to that of human enteric viruses.

The above points illustrate that these bacteriophages behave in many ways similar to human enteric viruses and thus would seem to be ideal as indicators of their presence in drinking waters. However, as with the Bifidobacteria, bacteriophages of *B. fragilis* are anaerobic, which involves complicated and tedious methodology which limits their suitability as routine indicator organisms (Tartera, Lucena and Jofre, 1989).

5.5 BIFIDOBACTERIA

Bifidobacteria are Gram-positive, non-sporulating, obligately anaerobic catalase-negative, rod-shaped bacteria. They are common in the faeces of humans, pigs, cattle and sheep, but not in horses, poultry or household pets such as cats and dogs. They were first described by Tissier in 1912. Over 17 species of *Bifidobacteria* are now recognized (Resnick and Levin, 1981) with the sorbitol-fermenting strains with *B. adolescentis* and *B. breve* only found in humans.

Bifidobacteria were first proposed as a possible indicator of faecal pollution by Mosel in 1958 because:

- They are anaerobic and cannot multiply outside the intestine (Dutka, 1979; Levin and Resnick, 1981).
- Bifidobacteria are exclusively faecal in origin, with some species occurring in humans in proportions exceeding that of *E. coli* (Resnick and Levin, 1981) (Table 5.1).
- They are considered ideal indicator organisms for tropical samples as such samples may contain organisms which can multiply, ferment lactose and produce indole at 44 °C, but are not faecal in origin (Carrillo, Estrada and Hazen, 1985; Evison and James, 1973; Feacham *et al.*, 1982).

The main drawback in using *Bifidobacterium* as an indicator organism is the lack of a suitable selective media and the difficulty with anaerobic methodology (Levin and Resnick, 1981). Various media have been developed in an attempt to overcome these restrictions (Evison and Morgan, 1978; Levin and Resnick, 1981; Mara and Oragui, 1983; Munoa and Panes, 1988), but for various reasons, including a lack of selectivity and the tendency of some of these media to inhibit certain *Bifidobacterium* species, these have proved relatively unsatisfactory. More recently, species-specific gene probes for *Bifidobacterium* species have been designed (Yamamato, Morotomi and Tanaka, 1992), but no further evaluation has been made of these.

In addition to problems of isolation there are also certain characteristics of *Bifidiobacterium* which may limit its usefulness as an indicator organism:

- It has a short survival time in the environment. The time lapse between when samples are collected and examined must be kept to a minimum and even then only 60–70% of samples can be recovered (Levin and Resnick, 1981).
- *Bifidiobacterium* is sensitive to chlorine which renders this organism ineffective for examining chlorinated waters (Levin and Resnick, 1981).

5.6 *RHODOCOCCUS* SPP.

Rhodococcus coprophilus is an aerobic, *Nocardia*-like actinomycete commonly found on herbivore dung and in aquatic environments (Rowbotham and Cross, 1977). As this organism has been found to be excreted only by farm animals, it has been suggested as a specific indicator organism of farm animal contamination. In a survey of a wide range of animals and birds, Mara and Oragui (1981) found that this organism was not recovered from faecal specimens of humans, cats, rabbits, rats, mice or turkeys. The frequency of isolation was 100% in cattle, sheep, pig and horse faeces. *Rhodococcus coprophilus* survives longer and better in aquatic environments than other indicator organisms such as faecal streptococci and *E. coli* (up to 100 days in polluted waters) and has been suggested as a useful indicator of the presence of remote faecal pollution of farm animal origin. The presence of both *Str. bovis* and *R. coprophilus* would indicate animal pollution of recent origin whereas the presence of *R. coprophilus* alone is suggestive of remote animal pollution. The use of this organism as an indicator organism is somewhat limited because of the long incubation period it requires (17–18 days), and the difficult isolation technique required (Mara and Oragui, 1981), factors that make it unpractical for routine monitoring. It is, however, useful in the assessment of the suitability of surface and ground water resources for supply purposes in rural areas that are at risk from farm animals.

5.7   HETEROTROPHIC PLATE COUNT BACTERIA

Heterotrophic plate counts (HPC) represent the aerobic and facultatively anaerobic bacteria that derive their carbon and energy from organic compounds. This group includes Gram-negative bacteria belonging to the following genera: *Pseudomonas, Aeromonas, Klebsiella, Flavobacterium, Enterobacter, Citrobacter, Serratia, Acinetobacter, Proteus, Alcaligenes* and *Moraxella* (Bitton, 1994). This already has been discussed in detail in section 1.5. Certain HPC organisms are considered to be opportunistic pathogens (section 6.4) and have been implemented in gastro-intestinal illness (Geldreich and Rice, 1987).

Heterotrophic bacteria such as those listed in Table 5.6 are commonly isolated from raw waters and are widespread in soil and vegetation. They can survive for long periods in water and rapidly multiply, especially at summer temperatures. In reality the counts themselves have little health significance, although there is evidence that excessive numbers can cause gastro-intestinal illness in the very young and other high-risk groups. There is also concern that these organisms can rapidly multiple in bottled waters, especially if not stored properly once opened (Gray, 1994). The EU Drinking Water Directive requires that there is no significant increase from background levels of HPC bacteria in either tap or bottled waters (EU, 1995). While HPC bacteria are not a direct indicator of faecal contamination, they do indicate variation in water quality and the potential for pathogen survival and regrowth.

Table 5.6 HPC bacteria found in distribution and raw waters (Bitton, 1994)

Organism	Distribution water		Raw water	
	Total	% of total	Total	% of total
Actinomycete	37	10.7	0	0
Arthrobacter spp.	8	2.3	2	1.3
Bacillus spp.	17	4.9	1	0.6
Corynebacterium spp.	31	8.9	3	1.9
Micrococcus luteus	12	3.5	5	3.2
Staphylococcus aureus	2	0.6	0	0
S. epidermidis	18	5.2	8	5.1
Acinetobacter spp.	19	5.5	17	10.8
Alcaligenes spp.	13	3.7	1	0.6
F. meningosepticum	7	2.0	0	0
Group IVe	4	1.2	0	0
Group M5	9	2.6	2	1.3
Group M4	8	2.3	2	1.3
Moraxella spp.	1	0.3	1	0.6
Pseudomonas alcaligenes	24	6.9	4	2.5
P. cepacia	4	1.2	0	0
P. fluorescens	2	0.6	0	0
P. mallei	5	1.4	0	0
P. maltophilia	4	1.2	9	5.7
Pseudomonas spp.	10	2.9	0	0
Aeromonas spp.	33	9.5	25	15.9
Citrobacter freundii	6	1.7	8	5.1
Enterobacter agglomerans	4	1.2	18	11.5
Escherichia coli	1	0.3	0	0
Yersinia enterocolitica	3	0.9	10	6.4
Group IIK biotype I	0	0	1	0.6
Hafnia alvei	0	0	9	5.7
Enterobacter aerogenes	0	0	1	0.6
Enterobacter cloacae	0	0	1	0.6
Klebsiella pneumoniae	0	0	0	0
Serratia liquefaciens	0	0	1	0.6
Unidentified	65	18.7	28	17.8
Total	347	100	157	99.7

Heterotrophic plate counts are done normally using the spread plate method using Yeast Extract Agar (YEA) and incubated at 22 °C for 72 hours and 37 °C for 24 hours. Results are expressed as colony forming units (cfu). Counts at 37 °C are especially useful as they can provide rapid information of possible contamination of water supplies. The low nutrient medium R2A agar (Reasoner and Geldrich, 1985) is recommended instead of YEA for the recovery of disinfectant-damaged bacteria (Department of the Environment, 1994a).

Heterotrophic plate counts have long been employed to evaluate

Table 5.7 Sanitary quality of water according to
Miguel* (Oliveri, 1982).

Quality	No. of bacteria/mL
Excessively pure	< 10
Very pure	10–100
Pure	100–1000
Mediocre	1000–10 000
Impure	10 000–100 000
Very impure	> 100 000

*Miguel, P. (1891) *Manuel Pratique d'Analyse Bacteriologique des Eaux*, Paris.

water quality. Miguel, in 1891, provided standards, based on HPCs, to evaluate water quality (Pipes, 1982a) (Table 5.7). In Britain, not much importance is placed on HPC for assessing the potability of drinking water. It is considered that their value lies mainly in indicating the efficiency of various water treatment processes including disinfection, as well as the cleanliness and integrity of the distribution system (Department of the Environment, Department of Health and Social Security, and Public Health Laboratory Service, 1983). Changes in the pattern of colony counts of samples taken from a given supply have more significance than a single numerical count. A sudden marked change in the colony count of water in a supply may indicate more serious pollution, whereas deviations in the expected seasonal trend may suggest longer-term changes in the water supply. In contrast, considerably more emphasis has been put on the importance of sampling and analysing HPC bacteria in the USA by the EPA, so much so that the National Primary Drinking Water Regulations now include maximum containment levels of no more than 500 cfu/mL for HPCs (US EPA, 1990b). This is to reduce possible interference with the detection of coliforms. While many HPC bacteria are recognized as opportunistic pathogens the full public health significance of such organisms to the general public is not yet known. Subsequently, it is likely that more emphasis will be placed on the value of such counts in the future.

5.8   OTHER INDICATOR ORGANISMS

There are a number of other organisms which have also been considered as having potential as alternative indicator organisms including *Pseudomonas* spp., *Bacteroides* spp. and *Candida albicans*.

   Pseudomonas aeriginosa is a Gram-negative, non-sporulating opportunistic pathogen which causes infection in wounds, as well as ear and urinary tract infections, meningitis and respiratory infections (Feacham et al., 1983). It has interesting growth properties, forming both oxidase and catalase, growing at 42 °C but not at 4 °C. It is also able to reduce nitrates and nitrites, produce ammonia from the breakdown of acetamide, and is able to hydrolyse casein but not starch. An important characteristic of the pseudomonad is that it can produce the blue-green pigment pyocyanin or the fluorescent pigment fluorescein, or both. *Pseudomonas aeriginosa* is particularly associated with disease among swimmers. Numerous cases of folliculitis, dermatitis, ear and urinary tract infections due to *P. aeriginosa*, contracted after swimming in contaminated waters, have been reported (Yoshpe-Purer and Golderman, 1987). Because of this association, and its consistent presence in high numbers in sewage, *P. aeriginosa* was thought to have potential as an indicator of water quality, particularly recreational waters (Cabelli, 1978). However, as this organism is known to be ubiquitous in nature and can multiply under natural conditions, it is in practice of little use in faecal contamination studies. While it should not be used as an indicator organism, bottled waters are required to be free of the organism and so must be monitored at bottling plants (EU, 1995). Brief details of the isolation of *P. aeriginosa* is given in section 1.5, with full details given by the Department of the Environment (1994a) and APHA (1992).

Bacteroides spp. are Gram-negative, non-motile, non-sporing, obligately anaerobic bacteria which form a major component of human and animal faeces (Table 5.1). In fact they are more numerous than *E. coli* in human faeces. There are currently five species: *B. diastonis, B. fragilis, B. ovatus, B. thetaiotaomicron* and *B. vulgatus. Bacteroides fragilis* is the species of the genus most commonly associated with human faeces. They are of limited value as indicator organisms mainly because of the problems associated with the isolation and enumeration of anaerobes. Allsop and Stickler (1984) report that in practice they appear to have little advantage over *E. coli*, especially as they rapidly die off in water.

Candida albicans is the most exciting potential indicator to be evaluated in recent years (Grabow, Burger and Nupen,1980). It is an extremely widespread yeast found in the populations of all developed countries, with up to 80% of the adults having low levels of infection and detectable levels of the yeast in their faeces. Although it lives at low levels of activity in the rectum of most people, it can result in mouth, vaginal, groin and general skin infections. The latter are especially common in swimmers (Buckley, 1971), where there has been a steady increase in the incidence of serious infections (Briscou, 1975). For most people infection becomes problematic in the intestinal tract after a course of antibiotic treatment. The yeast cannot exist for prolonged periods without a natural host and does not exist or grow independently in water, therefore the presence of *C. albicans* is a direct result of faecal contamination and makes it ideal for monitoring water, sewage and estuarine waters. The yeast is effectively removed by sewage treatment (Buck, 1977), although concentrations of up to 100 cells of *C. albicans* per 100 mL have been reported in treated effluents (Dutka, 1979).

The coliform index has been used for many years to determine the safety of swimming-pool water, yet the contamination was not always faecal in origin with infections primarily of the respiratory tract, skin and eyes (Esterman *et al.*, 1984; Robinton and Mood, 1966). For this reason *Staphylococcus aureus* and *C. albicans* have been proposed as better indicators of the types of infection associated with swimming (Sato *et al.*, 1995). There are no universally recognized standard enumeration procedures for *C. albicans*, although Dutka (1978) has evaluated a membrane filtration system and medium originally developed by Buck (1977) that he found to be both rapid and extremely reproducible. For example, typical counts for bathing beaches with high faecal coliform counts can be up to 25 °C. *albicans* per litre, falling to 0–2 *C. albicans* per litre for relatively unpolluted beaches. As an indicator of faecal contamination of drinking waters it has yet to be fully evaluated. Sato *et al.* (1995) have compared several methodologies and selective culture media for the isolation and enumeration of *C. albicans*. They found that the membrane filtration method using m-CA agar and incubated at 35 °C for 2–4 days (Buck and Bubucis, 1978) was most effective. Colonies are chocolate brown and approximately 1 mm in diameter. Confirmation of colonies was done by germ tube and chlamydospore production and sugar assimilation (Dutka, 1978).

5.9   ALTERNATIVE INDICATORS FOR TROPICAL ENVIRONMENTS

The use of total and faecal coliforms as a means of determining the microbiological quality of water continues to be recommended globally, despite the fact that it has been established that it is not feasible to condemn water supplies in tropical environments solely because of the presence these organisms (Chapter 2). A review by Toranzos (1991) of alternative indicators in tropical environments showed that the isolation of faecal streptococci appears to be as common as faecal coliforms in tropical waters which also negates their usefulness as indicators in these areas. The same problem arises with the use of *Cl. perfringens*. This organism also has the added disadvantage of requiring more complex methodology. The possibility of using *Bifidobacterium* has already been discussed but again being an anaerobic organism, enumeration of this organism requires complicated techniques.

The adequacy of bacteriophages as indicators in tropical waters is still being evaluated. In Puerto Rico, coliphages have been detected in contaminated, but not in pristine, waters. However, the use of coliphages assumes a direct correlation between concentrations of coliphages and concentrations of the bacterial indicators (i.e. total and faecal coliforms). This implies that the main concern behind the use of coliphages is the detection of coliforms, which in tropical areas has proved a pointless exercise.

It would appear that at present, there is no indicator organism available to adequately indicate faecal contamination of tropical water supplies. Therefore, it may be more appropriate to pursue a completely different approach to the problem, particularly as it is widely acknowledged that a major problem in water management in tropical areas has been the inappropriate transfer of Western standards (i.e. the coliform index) to such nations (McDonald and Kay, 1988). It is likely that in the future, water quality assessment in tropical nations will be based on a risk assessment approach, taking into account appropriate social, political, environmental and economic factors. Such an approach would allow the establishment of water quality assessment more in tune with the limitations of indicator organisms and methods of detection currently in use.

5.10 CHEMICAL INDICATORS

FAECAL STEROLS

Dutka, Chan and Coburn (1974) have proposed using faecal sterols such as coprostanol and cholesterol to indicate human contamination of water systems. Both of these sterols are present in mammalian faeces and have been found in domestic sewage and receiving wastes. Unlike biological indicators, faecal sterols do not seem to be affected by chemical disinfectants or toxic wastes discharges. Thus, for chlorinated effluents, faecal sterol levels would appear to be ideal indicators of human faecal contamination and possibly of health hazards associated with non-activated viruses. The use of faecal sterols may be of particular benefit in situations where the use of conventional bacterial indicators is difficult, such as industrial effluents or in situations where it can not be established whether the bacterial indicators are of faecal origin or are land washed (Dutka, 1979; Grimalt et al., 1990).

Humans and higher animals excrete high concentrations of cholesterol, a precursor to coprostanol, in their faeces. It is also found in some non-faecal sources (Waite, 1984). This lack of specificity means that this particular sterol is not given serious consideration as a potential indicator. On the other hand, coprostanol is stable and non-pathogenic.

It is easily degraded by sewage treatment and therefore its presence in water is indicative of recent sewage contamination (Waite, 1984). Sewage effluents contain on average 33 g/L of coprostanol of which between 80–95% is associated with particulate matter and quickly settles out of solution to become assimilated into the aquatic sediment (Brown and Wade, 1984). For this reason it is becoming increasingly common to measure recent or intermittent faecal pollution of surface and coastal waters by the examination of coprostanol in the sediment (Hasset and Lee, 1977; Walker, Wun and Litsky, 1982; Writer et al., 1995).

Dutka and El-Shaarawi (1975) found that there is no consistent relationship between bacterial parameters and faecal sterols but that they could still be considered unequivocal indicators of the presence of faecal contamination. Problems associated with their use as faecal contamination indicators include the laborious time required to process samples. Monitoring coprostanol requires an elaborate extraction process with chromatography which is not feasible on a routine basis. Also, there is, at present, a lack of information regarding background levels of faecal sterols in the natural environment.

UROBILINS

The most recent suggestion for alternative indicators of faecal pollution are urobilins. These compounds are formed from conjugated bilirubin by hydrolysis and reduction by intestinal microflora and thus originate only from mammalian faeces and urine (Miyabara et al., 1994). To date, they have been used to detect hepatic function and have only recently been suggested as a potential indicator. Studies in Japan indicate that they remain relatively stable in river water (Miyabara et al., 1994). Table 5.8 shows how urobilins correlate with routinely used indicators.

Urobilins are detected using high performance liquid chromatography (HPLC) with fluorometry based on the Jaffe Schlesinger reaction. At

Table 5.8 Correlation coefficients (r) and significance (p) between i-urobilin and other indicators of faecal pollution (Miyabara et al., 1994)

COD	0.02	$p > 0.05$
Ammonia nitrogen	0.72	$p < 0.01$
Total coliform	0.83	$p < 0.001$
Faecal coliform	0.81	$p < 0.001$
n	15	

present, due to the limited research into these compounds, it is difficult to assess their potential as indicator organisms.

5.11 CONCLUSIONS

The primary conclusion is that there is no universal indicator for water-borne pathogens and that each kind of pollution should be measured by the method (or organism) most appropriate to it. According to Mossel (1982), two distinctly different functions of a model indicator organism can be distinguished, the *index* and *indicator* functions. Index organisms relate in some way to the health risks or risk of given pathogens. The criteria which should be met by the ideal index organism are very similar to those required by traditional faecal indicators. Indicator organisms have a more general function and relate to the efficiency of a given treatment procedure. For instance, the use of plate counts to evaluate swimming-pool disinfection. Clearly, a given organism may fulfil both index and indicator function roles. This concept highlights the fact that different organisms may be required to characterize different pollution sources. All too frequently indicator systems are used to obtain information beyond their capabilities or the data obtained is interpreted without due regard to the limitations of the particular organism used. In the end, the best indicator will be the one whose densities correlate best with the health hazards associated with a given type of pollution (Cabelli, 1978).

There are other factors which must also be taken into account when deciding the best system to use, such as the complexity of the procedure selected and the cost. Many proposed indicator organisms involve complex and expensive methodologies which are beyond the means and capabilities of smaller utilities and developing countries, and therefore could not be applied on any sort of a routine basis. It would therefore appear that currently used bacterial indicators will go on being used throughout the world until an alternative is found that meets all the criteria established for indicator organisms and that is acceptable in terms of expense and simplicity.

The significance of emerging pathogens on water quality assessment

6

The use of indicator organisms is the barometer by which the safety of water for human consumption is judged. However, in recent years there has been an increase in the numbers of incidents where water-borne outbreaks have occurred in waters which met the required standards for indicator organisms, in particular the coliform index. Many of these outbreaks have been caused by viral and protozoan agents, particularly the enteric viruses, and the protozoa *Giardia* and *Cryptosporidium*. This chapter looks at these organisms, as well as some of the more traditional bacterial pathogens, and examines the significance that their increasing frequency has for water quality assessment.

6.1 GIARDIA LAMBLIA

Giardia lamblia is a flagellated protozoan belonging to the class Zomastigophorasida, order Diplomonadonda and family Hexamitidae. It was first discovered in 1681 by Leeumenhoek who found the organism in his own stools and was later named *Giardia lamblia* after Giard who studied the parasite and Lamb who first described it. For more than a century after its initial discovery, the pathogenic potential of *Giardia* was not fully appreciated. Indeed, up to the 1960s, it was widely believed to be a commensal parasite of doubtful pathogenicity. It is now recognized that this organism is a significant cause of gastro-enteritis ranging from mild to severe and debilitating disease with a worldwide distribution (Akin and Jakubowski, 1986).

Giardia lamblia exists in a trophozoite and cyst form. Trophozoites are easily recognized. Their bodies are pear or kite shaped, approximately 9–21 μm long by 6 μm wide with an anterior sucking disc on the flattened ventral surface (Figure 6.1). There are four pairs of flagella and the organism is binucleate (Feacham *et al.*, 1983). Cysts are ovoid, 14–16 μm long and 6–12 μm wide and are quadrinucleate. *Giardia* cysts

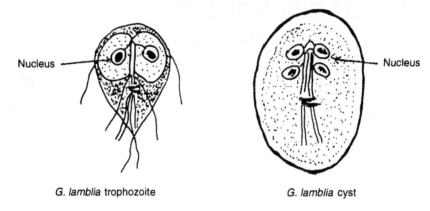

Nucleus

Nucleus

G. lamblia trophozoite   *G. lamblia* cyst

Figure 6.1   Trophozoite and cyst of *Giardia lamblia* (Lin, 1985).

are relatively resistant to environmental conditions and are capable of survival once excreted for long periods, especially in winter.

The life cycle of *G. lamblia* is discussed in considerable detail by Lin (1985). In brief, the life cycle begins when an ingested cyst passes into the stomach of an exposed individual. Excystation follows with the resulting trophozoite attaching itself to the epithelium of the small intestine by an adhesive disk. The trophozoite multiplies by binary fission to large numbers which move down the intestinal tract transforming into cysts in the ileum. The cysts are passed from the body in the stools (Figure 6.2) (Akin and Jakubowski, 1986).

The symptoms of giardiasis, or backpackers' disease as it is commonly known in the USA due to its high incidence amongst those who drink unfiltered and non-disinfected water from mountain streams, develops between 1–4 weeks after infection. The predominant clinical feature associated with giardiasis is severe diarrhoea, occurring in 50% of those infected by the pathogen. Other symptoms range from malaise, weakness, fatigue, dehydration, weight loss, distension, flatulence, anorexia, cramp-like abdominal pain, and epigastric tenderness to steatorrhea and malabsorption (Lin, 1985). Generally giardiasis occurs only as a mild disease but it can develop into a serious illness. It has a worldwide distribution with a prevalence of about 7% and is three times more common in children than in adults (Ellis, 1989). Asymptomatic giardiasis may also occur (Akin and Jakubowski, 1986). Usually only the symptomatic patient is treated and there are a number of drugs available for this (Lin, 1985).

The exact infectious dose necessary for infection is not exactly known but it is thought to be somewhere between 25 and 100 cysts (Feacham *et al.*, 1983; Ellis, 1989).

Transmission of *Giardia* cysts may be by faecal contamination of

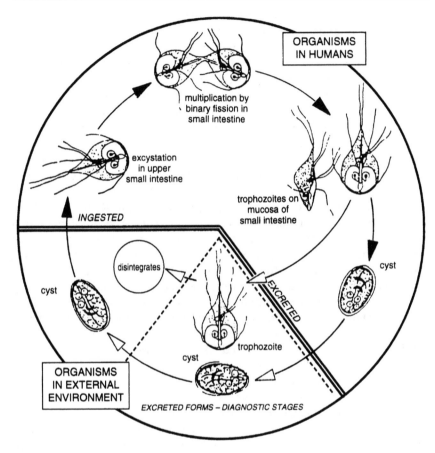

Figure 6.2 Life cycle of *Giardia lamblia*.

either hands, food or water supplies (Feacham *et al.*, 1983). Originally cysts were thought to be host specific. However, it is now well established that the disease is a zoonosis. Cysts from a human source can be infective to about 40 different species of animal including guinea-pigs, dogs, beavers, racoons and big horn montflou sheep, and *vice versa* (Faubert, 1988). Cysts are therefore widely distributed in the environment entering waterways through sewage or storm water discharges or via the droppings of infected animals. A study of raw water supplies in the USA by Le Chevallier, Norton and Lee (1991a) found that *Giardia* cysts are present in as many as 81% of raw water supplies largely due to the introduction of sewage effluents and in 17% of finished water supplies (Le Chevallier, Norton and Lee, 1991b). A similar survey in Scotland found that 48% of raw waters and 23% of treated water supplies sampled contained cysts (Gray, 1994). Waterborne transmission of giardiasis is well documented (Lin, 1985; Akin and Jakubowski, 1986) and is now established to be one of the most common causes of water-borne diseases in the developed world. It is particularly common in the USA where it is now considered to be endemic with a carrier rate of 15–20% of the population, depending on their socio-economic status, age and location. *Giardia* is the most common animal parasite of humans in the developed world, although water is probably not the most common mode of transmission. However, *Giardia* remains one of the most common causes of waterborne diseases. In the USA, *Giardia lamblia* causes more waterborne disease than any other identifiable aetiologic agent with more than 697 outbreaks reported between 1989 and 1990 (Herwaldt *et al.*, 1992). The first major documented outbreak of giardiasis associated with water supply occurred in Rome, New York (USA), with approximately 50 000 people affected. The outbreak was a result of consumption of water that had been chlorinated but not filtered. The number of reported cases in England and Wales has risen from 1000 per annum in the late 1960s to over 5000 per annum by the late 1980s, although these were largely associated with people travelling overseas. The number of outbreaks in the UK associated with drinking water contamination is steadily increasing with the most significant outbreak of giardiasis occurring in South Bristol (UK) in the summer of 1985 when 108 cases were reported (Browning and Ives, 1987; Jephcote, Begg and Baker, 1986). It is thought that contamination of the water supply occurred in the distribution system and was not due to any failure of the water treatment process.

Craun (1977) indicated that most *Giardia* outbreaks occur in waters where chlorination is the only form of water treatment. This resistance to disinfection levels typically used in water treatment indicates the need for additional treatment barriers. In recognition of this, recent amendments to the US Safe Drinking Water Act now require that all surface waters intended for human consumption must undergo filtration to specifically remove cysts, and sufficient disinfection to destroy *Giardia* and prevent disease transmission. To date no such provisions exist in European or British legislation. There is no way of preventing infection except by adequate water treatment and resource protection. Current disinfection practices are generally inadequate as the sole barrier to prevent outbreaks. Boiling water for 20 minutes will kill cysts, while the use of 1 μm pore cartridge filters to treat drinking water at the point of use are also effective.

Methods for detecting *Giardia* cysts have been available since about 1975, however, none to date have been especially successful. At present, the method most commonly used is that outlined in Standard Methods (Figure 6.3) (APHA, 1992). A similar method is recommended in Britain (Standing Committee of Analysts, 1990). *Giardia* is not free living and therefore is unlikely to reproduce outside the host animal. Because of this, cysts are generally present in very low densities and so samples must be concentrated first. This is achieved by using ultra filtration cassettes or finely wound polypropylene cartridges (APHA, 1992). As cyst numbers cannot be amplified by *in vitro* cultivation, they are generally detected by immunofluorescence with poly- or monoclonal antibodies or by direct phase contrast microscopy (Sauch, 1986;

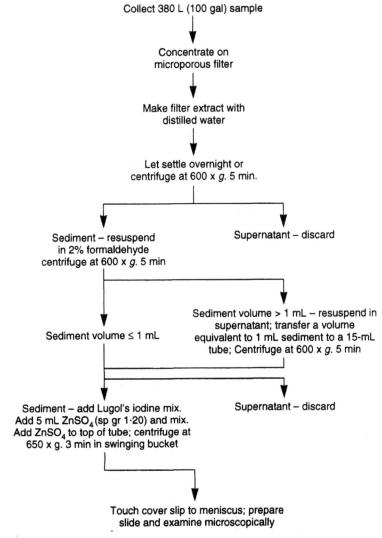

Figure 6.3   Method for isolating *Giardia lamblia* (APHA, 1992).

Standing Committee of Analysts, 1990). More recently a cDNA probe has been constructed for the detection of *Giardia* cysts in water and wastewater (Abbazadegan, Gerba and Rose, 1991). The main problem with these is that they do not allow for differentiation between viable and non-viable cells. Mahbubani *et al.* (1991) have now developed a technique using PCR which allows some distinction between dead and viable cells.

6.2 *CRYPTOSPORIDIUM* SPP.

Cryptosporidium spp. are coccidian protozoan parasites which belong to the phylum Apicomplexa (Barer and Wright, 1990). At least four species are recognized: two in mammals (*C. parvum* and *C. muris*) and two in birds (*C. baileyi* and *C. meleagridis*). *Cryptosporidium parvum* is the species primarily responsible for clinical illness in humans and animals (Rose, 1988). Despite its discovery early in 1912, the first case of human infection was not recorded until 1976 (Gray, 1994). *Cryptosporidium* is now recognized as being a significant cause of gastro-intestinal disease in human beings, particularly in children and immuno-compromised individuals.

Detailed reviews of the biology and life cycle of *Cryptosporidium parvum* can be found in Rose (1988) and in Fayer and Ungar (1986). In brief, the parasite has a complex life cycle involving both sexual and asexual stages (Figure 6.4). Following ingestion, the infectious stage, i.e. the oocyst, releases four sporozoites after excystation. The sporozoite differentiates into the trophozoite which undergoes asexual multiplication to form Type 1 meronts and then merozoites. It is during asexual reproduction that recycling of the merozoite stage occurs. The sexual stage occurs when merozoites form Type 2 meronts and produce microgametes and macrogametes, which fuse and form the zygote which is then excreted in the faeces (Rose, 1988). About 80% of zygotes produced are thick-walled oocysts which are immediately infectious when excreted. These oocysts are very resistant and are capable of survival outside the intestine for considerable lengths of time (Smith and Rose, 1990). The remaining 20% are thin walled and are responsible for auto-infection. This permits *Cryptosporidium* to complete its life cycle within a single host, so once infected the host becomes a life-time carrier and subject to relapses ensuring that high numbers of oocysts are continually being produced (Fayer and Ungar, 1986).

Cryptosporidiosis is acquired by ingesting viable oocysts (Barer and Wright, 1990). In immuno-competent individuals cryptosporidiosis is a common cause of acute self-limiting gastro-enteritis with symptoms commencing on average 3–4 days after infection and lasting for up to two weeks. Clinical symptoms include an influenza-like illness,

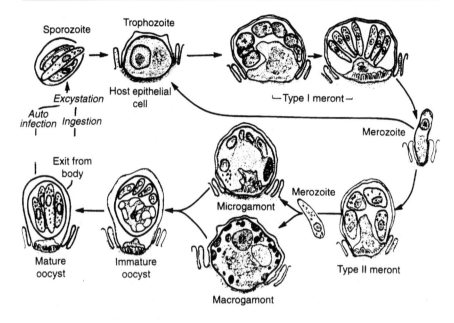

Figure 6.4 Life cycle of *Cryptosporidium parvum* (Fayer and Ungar, 1986).

diarrhoea, malaise, abdominal pain, anorexia, nausea, flatulence, malabsorption, vomiting, mild fever and weight loss. Generally this disease is not fatal among healthy individuals. However, in young malnourished children it can cause severe dehydration and sometimes death. Infection in the immuno-competent has been described worldwide with a prevalence of 0.6–20% reported in Western countries and 4–20% in developing countries (Smith and Rose, 1990).

In the immuno-compromised, including those with AIDS, other acquired abnormalities of T lymphocytes, congenital hypogammaglobulinaemia, severe combined immunodeficiency syndrome, those receiving immuno-suppressive drugs and those with severe malnutrition, cryptosporidiosis can become a life-threatening condition causing profuse intractable diarrhoea with severe dehydration, malabsorption and wasting. Sometimes the disease spreads to other organs. These symptoms can persist unabated until the patient eventually dies. Among AIDS patients cryptosporidiosis has a prevalence of 3–4% in the USA and greater than 50% in Africa and Haiti (Smith and Rose, 1990). A study on the filtration of drinking water in California (USA) has shown that AIDS suffers are no more at risk of contracting cryptosporidiosis than the rest of the population from the presence of the pathogen in drinking water, the higher incidence being linked to sexual transmission of the disease between homosexual and bisexual men

(Sorvillo *et al.*, 1994). However, the benefit of filtration of drinking water to prevent outbreaks of the disease is well established, especially during periods of heavy *Cryptosporidium* contamination. In the UK, cryptosporidiosis is currently the fourth most commonly identified cause of diarrhoea in which a parasitic, bacterial or viral cause was established. Approximately 2% of all current cases of diarrhoea in the UK are due to *Cryptosporidium*, with children more at risk than adults (Department of the Environment and Department of Health, 1990).

At present there are no effective drugs to control or treat the infection although there has been reported success using passive chemotherapy (WRc, 1991). Little is known of the exact infectious dose size, but it is thought to be small, probably less than 10 oocysts (Casemore, 1990). Recent studies indicate that a single oocyst may be enough to cause infection (Blewett *et al.*, 1993), although outbreaks of cryptosporidiosis have been associated with gross contamination (Wilkins, 1993).

At first cryptosporidiosis was thought to be a zoonosis contracted by direct contact with animals, especially young farm animals and their faeces. The organism is not host specific and is capable of infecting many species of mammal, bird and reptile (Packham, 1990). The disease is now known to readily spread from person to person, especially young children (Department of the Environment and Department of Health, 1990). Oocysts from humans are infective for numerous mammals (Fayer and Ungar, 1986). Both foodborne and airborne routes have been documented, but further evidence is required to clarify the significance of both these transmission routes. Water has increasingly been recognized as an important indirect route of transmission and Smith and Rose (1990) have identified certain factors which favour waterborne cryptosporidiosis:

- lack of species specificity;
- close association between human and animal hosts;
- low infective dose;
- monoxenous development, with autoinfective cycle;
- large numbers of oocysts excreted;
- fully sporulated oocysts excreted;
- oocysts are environmentally robust and chlorine insensitive;
- small size of oocysts.

Waterborne transmission of cryptosporidiosis was first suggested because of its association with travellers' diarrhoea and with *Giardia* infections, both of which are well documented causes of waterborne diarrhoea (Smith and Rose, 1990). A number of surveys of the incidence of cryptosporidiosis were carried out in the 1980s. During this period, a number of outbreaks were reported. In some of these incidents oocysts were detected in the water supply of those affected, while in others there was strong circumstantial evidence that water was the vehicle of transmission (Barer and Wright, 1990). Studies of water resources in the UK and the USA found that oocysts commonly occurred in all types of surface water (lakes, reservoirs, streams and rivers) including pristine waters with densities ranging from 0.006–2.5 oocysts/litre (Badenoch, 1990; Le Chevallier, Norton and Lee, 1991a; Madore *et al.*, 1987). Significantly higher numbers of oocysts are found in water resources receiving untreated or treated wastewaters, while oocysts occur much less frequently in groundwaters. These findings suggest that a background level of oocysts exists in many waters which may suddenly become increased by accidental pollution or by heavy pollution, especially if such situations follow shortly after solid or liquid manure has been added to land adjacent to watercourses (Department of the Environment and Department of Health, 1990). The Badenoch report (Department of the Environment and Department of Health, 1990) also found that it is probable that most of the oocysts found in both surface and groundwaters are derived form agricultural sources. Oocyst contamination can occur via several routes (Department of the Environment and Department of Health):

- slurry stores may leak or burst and overflow;
- seepage from solid manure stores;
- run-off from soiled yards;
- direct contamination from grazing animals near water courses;
- manure may percolate through soil to field drainage systems and ultimately to water courses;
- disposal of sewage treatment processes containing contaminated sewage.

Oocysts are ovoid, are between 4–6 μm in diameter and tend to occur in low numbers in water (Barer and Wright, 1990). Their detection in water samples relies on filtration of large volumes of water to remove oocysts and examination of the concentrate by microscopy. Most methods available for oocyst detection are adaptations of those used for *Giardia* detection (Musial *et al.*, 1987; Ongerth and Stibbs, 1987; Standing Committee of Analysts, 1990). Current standard recovery methods involve the passage of large volumes of water (100–500 litres), at rates of approximately 1.5 litres per minute, through wound propylene fibres with a 1 μm pore size (Musial *et al.*, 1987; Standing Committee of Analysts, 1990). Polycarbonate filters are also commonly used (Ongerth and Stibbs, 1987). The resulting concentrate is eluted and the oocysts are counted. Varying recovery rates have been reported.

Identification of *Cryptosporidium* oocysts depends on determination of their size, shape and staining characteristics. Two approaches are available: direct staining remains (such as modified Ziehl–Neelsen) or the use of fluorescent labelled antibodies (IFAT) which bind specifically to the surface of the oocyst. However, none of the available identification methods distinguish between *Cryptosporidium parvum* oocysts and the oocysts of other species of *Cryptosporidium*, nor do they indicate whether the oocysts are viable (Hayes and Cooper, 1994).

A survey of a number of surface water filter plants by Le Chevalier, Norton and Lee (1991a) demonstrated that *Cryptosporidium* oocysts are frequently isolated from filtered drinking waters (approximately 27%). Finding oocysts in water does not necessarily mean that the population is at risk. However, because the minimum dose for cryptosporidiosis is thought to be so low and because of the fact that oocysts can withstand considerable environmental pressures, remaining viable for long periods of time at low temperatures, low-level contamination of a potable water supply has the potential to result in large scale infection of the population (Smith, 1992). Table 6.1 shows that many outbreaks have occurred in waters which have undergone water treatment. However, it is often difficult to identify the source of cryptosporidiosis contamination of water supplies (Maguire *et al.*, 1995). Since 1988 a number of major outbreaks of waterborne cryptosporidiosis have been recorded in the UK (Table 6.1). The first in 1988 occurred in Ayrshire when 27 cases were confirmed, although many more people were thought to have been infected. Of those infected, 63% were less than 8 years of age. The cause was thought to have been due to the finished water being contaminated from run-off from surrounding fields on which slurry had been spread. The oocysts were found in the chlorinated water in the absence of faecal indicator bacteria. The second outbreak was far more serious and affected over 400 people in the Oxford and Swindon areas early in 1989 (Richardson, Stuart and Wolfe, 1991), although according to Rose (1990) as many as 5000 people may have been affected. The outbreak was quickly traced to the Farmoor Water Treatment Plant near Oxford, which takes its water from the River Thames. On investigation oocysts were found in the filters, in the filter backwash water and in the treated water, even though it had been chlorinated and the microbiological tests had shown the water to be of excellent quality. The oocysts were found in the Farmoor Reservoir and also in a tributary of the River Thames, upstream of the treatment plant. The seasonal presence of the organisms was particularly associated with the grazing of lambs and with the scouring that often occurs in their early lives. While the rapid sand filters at Farmoor were removing 79% of the oocysts after coagulation, the recycling of the backwash water had given rise to exceptionally high concentrations of oocysts (10 000 per litre) with resultant breakthrough. Disinfection with chlorine was not effective, and the first action by Thames Water was to stop recycling the backwash water (up to then a normal practice) which brought the problem under control. There was a decrease in the number of reported cases as the number of oocysts in the water decreased. These oocysts can survive for up to 18 months depending on the temperature.

There have been further outbreaks since then, but on a much smaller scale (Table 6.1). One was in North-West Surrey at about the same time as the Oxford outbreak. Again water was taken from the River Thames downstream of Oxford. In April 1989, an outbreak occurred in Great Yarmouth following a change in the water supply, while in Hull there were 140 cases reported in January 1990. There was also an outbreak in the Loch Lomond area in 1989. In this case livestock grazing around the reservoir or in the catchment of rivers were the source of the organism. Cattle and infected humans can excrete up to 10^{10} oocysts during the course of infection, so that cattle slurry, wastewater from marts and sewage should all be considered potential sources of the pathogen. In Devon outbreaks of cryptosporidiosis have occurred in water supplied from the Littlehampton Water Treatment Plant regularly each year since 1991. In the USA there have been a number of reported outbreaks (Table 6.1). The most important outbreak of crypto-sporidiosis in the USA in recent years occurred in April 1993 in Milwaukee. The water distribution system serving 800 000 people was contaminated by raw water from a river swollen by spring run-off. In all 370 000 people became ill, 4400 were admitted to hospital and approximately 40 died (Jones, 1994).

These outbreaks suggest that present methods of water treatment are not adequate to deal with the parasite. Procedures of water treatment vary from place to place but all ultimately terminate with chlorination. It has been established that chlorination at levels used in water treatment is ineffective against oocysts, with concentrations as high as 16 000 mg/L required to prevent excystation (Korich *et al.*, 1990). Alter-

Table 6.1 Examples of outbreaks of cryptosporidiosis

Year	Country/location	Nos. affected	Suspected cause
1983	Surrey (UK)	16	Source contamination
1984	Texas (USA)	2000	Sewerage contamination of well
1985	Surrey (UK)	50	Source contamination
1987	Georgia (USA)	13 000	Operation irregularities
1988	Ayrshire (UK)	27	Post-treatment contamination
1989	Swindon/Oxfordshire (UK)	500	Source contamination
1989	L. Lomond, Scotland (UK)	442	Source contamination
1990	Humberside (UK)	140	Source contamination
1991	South London (UK)	44	Source contamination
1993	Milwaukee (USA)	370 000	Post-treatment contamination

native disinfectants for water treatment and inactivation of *Cryptosporidium* oocysts have been investigated including chlorine dioxide, monochloramine, hydrogen peroxide and ozone (Korich *et al.*, 1990; Peeters, van Opdenbosch and Glorieux, 1989). Of these, ozone is considered to have the most potential. Korich *et al.* (1990) found more than 90% inactivation after treating oocysts with 1 ppm of ozone for 5 minutes (Figure 6.5). Similar results were also observed by Parker, Greaves and Smith (1993) (Figure 6.6). The use of ozone is not without its disadvantages, in particular its organic by-products (Department of the Environment and Department of Health, 1990). A more recent report by the WRc (1994) does not consider ozone alone to be a practical or cost-effective method of protection against *Cryptosporidium* oocysts as very high doses would be required to achieve a high degree of oocyst inactivation. In any case it is considered that well-operated treatment processes with proper filtration and disposal or re-use of filter backwash water should be capable of achieving 99% reduction in oocyst concentration (West, 1991; WRc, 1994).

At present, little can be inferred about the safety or otherwise to the public if oocysts are detected in water samples. This is because of the many gaps in our knowledge of this organism, including their occurrence in natural waters, the minimum infective dose and their survival capabilities. The main problem with *Cryptosporidium* is the inability to

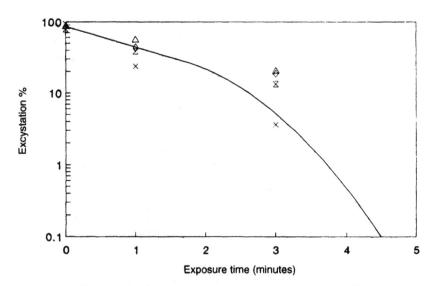

Figure 6.5 Decline in mean percent excystation of *Cryptosporidium* oocysts exposed to 1 ppm of ozone at 25 °C (Korich *et al.*, 1990).

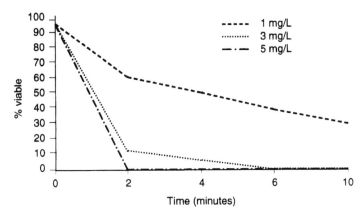

Figure 6.6 The effect of residual ozone concentrations on the viability of C. *parvum* oocysts at 20 °C (Parker, Greaves and Smith, 1993).

accurately detect viable oocysts, and therefore at present oocyst numbers are largely underestimated (Smith, 1992).

Protozoan pathogens of humans are almost exclusively confined to tropical and sub-tropical areas, which is why the increased occurrence of *Cryptosporidium* and *Giardia* cysts in temperate areas is causing so much concern. However, with the increase in travel, carriers of all diseases are now found worldwide, and cysts of all the major protozoan pathogens occur in European sewages from time to time. Two other protozoan parasites which occur occasionally in the UK are *Entamoeba histolytica* which causes amoebic dysentery and *Naegteria fowleri* which causes the fatal disease amoebic meningo-encephalitis.

Entamoeba histolytica is carried by about 10% of the population in Europe and 12% in the USA. In countries where the disease is widespread (not Europe) the highest rate of incidence occurs in those groups with unprotected water supplies, inadequate waste disposal facilities and poor personal hygiene. In Europe the number of cysts in sewage is generally low, less than five per litre in untreated raw sewage. Because of their poor settleability, like all protozoan cysts they tend to pass through the sewage treatment plant and reach surface waters. Where surface waters are reused for supply purposes, any cysts present will be taken up with the water. The disease amoebic dysentery is very debilitating, with the infection centred in the large intestine where the amoebae multiply and adhere to enterocytes, causing ulceration. The symptoms include mid-abdominal pain, diarrhoea alternating with constipation, or chronic watery diarrhoea with a discharge of mucus and blood.

The cysts of *Naegteria fowleri* normally enter the body through the nasal cavities while swimming in infected water although it is thought infection is possible through drinking contaminated water. Once in the body the protozoa rapidly migrate to the brain, cerebrospinal fluid and bloodstream. Cases are very rare but generally fatal, and are exclusively associated with swimming in warm contaminated waters, especially hot springs and spa waters. Since it was first reported in 1965 there have only been six reported cases in the UK. The last reported case was in 1978, an 11-year-old girl who was infected while swimming in a municipal swimming pool filled with water from a natural hot spring in the City of Bath.

6.3   VIRUSES

There are over 120 distinct known types of human pathogenic viruses. Of most concern for drinking water are those which cause gastro-intestinal illness, known as the enteric viruses, which include enteroviruses, rotaviruses, astroviruses, caliciviruses, Hepatitis A virus, Norwalk virus and other 'small round' viruses (West, 1991). As enteric viruses are usually gut-associated, the illnesses they cause are primarily gastro-intestinal in nature (Sellwood and Dadswell, 1991). However, the health risks presented by these viruses are not just restricted to gastro-enteritis (Table 6.2). Many of the enteroviruses such as Reovirus, Coxsackievirus and Echovirus, cause respiratory infections and are present in the faeces of infected people. Poliovirus in particular is common in sewage due to the vaccination programme within communities, but does not indicate actual infection. In fact virological examination of sewage has been used to document the effect of vaccination campaigns as those who are vaccinated with the live attenuated oral poliovirus vaccine shed faecal virus for a considerable time afterwards (Pöyry, Stenvik and Hovi, 1988). However, during outbreaks, wild poliovirus will also be identified in sewage and in waters receiving treated and untreated sewage (Avoort et al., 1995). Adenovirus 3 is commonly associated with swimming pools and can also cause pharyngo-conjunctival fever. In addition to those listed in Table 6.2, Coxsackie B virus has been associated with myalgic encephalomyelitis (ME), acute myopericarditis and dilated cardiomyopathy, a chronic cardiac disease which is the second most common reason for cardiac transplants in the UK (Watkins and Cameron, 1991).

Human viruses present in sewage are almost entirely derived from faecal matter. Viral contamination arises when sewage containing pathogenic viruses contaminates surface and ground waters which are subsequently used as sources of drinking waters (West, 1991). Large outbreaks of viral disease occur when massive sewage contamination takes place overwhelming existing water treatment mechanisms. Infectious hepatitis, enteroviruses, Reovirus and Adenovirus are all thought

Table 6.2 Some human enteric viruses and the diseases they cause (Bitton, 1994)

Virus group	Serotypes	Some diseases caused
Enteroviruses		
Polioviruses	3	Paralysis, aseptic meningitis
Coxsackievirus		
A	23	Herpangia, aseptic meningitis, respiratory illness, paralysis, fever
B	6	Pleurodynia, aseptic meningitis, pericarditis, congenital heart disease anomalies, nephritis, fever
Echovirus	34	Respiratory infection, aseptic meningitis, diarrhoea, pericarditis, myocarditis, fever and rash
Enteroviruses (68–71)	4	Meningitis, respiratory illness
Hepatitis A virus (HAV)		Infectious hepatitis
Reoviruses	3	Respiratory disease
Rotaviruses	4	Gastro-enteritis
Adenoviruses	41	Respiratory disease, acute conjunctivitis, gastro-enteritis
Norwalk agent (calcivirus)	1	Gastro-enteritis
Astroviruses	5	Gastro-enteritis

to be transmitted via water. Of most concern in Britain is viral hepatitis. There are three subgroups, Hepatitis A which is transmitted by water, Hepatitis B which is spread by personal contact or inoculation and which is endemic in certain countries such as Greece, and Hepatitis C which is a non A or B type Hepatitis virus. Hepatitis A is spread by faecal contamination of food, drinking water and areas which are used for bathing and swimming. Epidemics have been linked to all these sources, and it appears that swimming pools and coastal areas used for bathing which receive large quantities of sewage are particular sources of infection. The virus cannot be cultivated *in vitro* so studies are confined to actual outbreaks of the disease. Hepatitis A outbreaks usually occur in a cyclic pattern within the community as, once infected, the population is immune to further infection by the virus. So no new cases occur for 5 to 10 years until there is a new generation (mainly of children) which has not been previously exposed. There is no treatment for Hepatitis A, with the only effective protection good personal hygiene, and the proper protection and treatment of drinking water. Immuno globulin is often given to prevent the illness developing in possible contacts, although it is not always successful. Symptoms develop 15 to 45 days after exposure and include nausea, vomiting, muscle ache and jaundice. Hepatitis A virus accounts for 87% of all viral waterborne disease outbreaks in the USA (Craun, 1986). In June 1979, a large waterborne outbreak of gastro-enteritis and hepatitis occurred in Georgetown, Texas affecting approximately 79% of individuals supplied by the contaminated water following a period of heavy rainfall that washed sewage into the ground water supply. The best documented outbreak of waterborne viral disease occurred in New Delhi, India in 1955/56 when 35 000 cases of infectious hepatitis were reported following gross contamination of the water supply by sewage (Dennis, 1959).

Warm-blooded animals appear able to carry viruses pathogenic to humans. For example 10% of Beagles have been shown to carry human enteric viruses. Therefore there appears a danger of infection from waters not contaminated by sewage but by other sources of pollution, especially storm water from paved areas. Most viruses are able to remain viable for several weeks in water at low temperatures, so long as there is some organic matter present. Viruses are found in both surface and groundwater sources. In the USA as many as 20% of all wells and boreholes have been found to be contaminated with viruses. Two viruses which have caused recent outbreaks of illness due to drinking water contamination are Norwalk virus and Rotavirus (Craun, 1991; Cubitt, 1991).

Norwalk virus results in severe diarrhoea and vomiting. It is of particular worry to the water industry in that it appears not to be affected by normal chlorination levels. Also it seems that infection by the virus only gives rise to short-term immunity while lifelong immunity is conferred by most other enteric viruses. In 1986, some 7000 people who stayed at a skiing resort in Scotland became infected with a Norwalk-like virus. The private water supply, which was untreated, came from a stream subject to contamination from a septic tank. Table 6.3 lists some examples of Norwalk associated waterborne outbreaks. The largest of these outbreaks occurred in Rome, Georgia (USA) in 1980, when contaminated water from a textile factory came into contact with a community water supply. The largest viral associated outbreak of gastro-enteritis in Britain occurred in Branham in 1980 when over 3000

Table 6.3 Examples of waterborne outbreaks due to the Norwalk virus (Bitton, 1994)

Year	Location	No. reported ill	Remarks
1978	Tacoma, WA	600	Contaminated well
1979	Arcota, CA	30	Contaminated sprinkler system
1980	Maryland	126	Contaminated well
1980	Rome, GA	1500	Contaminated community water supply
1982	Tate, GA	500	Contaminated well and spring

cases were reported. This incident occurred when source boreholes became contaminated by sewage (Short, 1988).

Rotavirus is a major contributor to child diarrhoea syndrome. This causes the death of some six million children in developing countries each year. This is not, thankfully, a serious problem in Europe due to better hygiene, nutrition and health care. Outbreaks do occur occasionally in hospitals and, although associated with child diarrhoea, can be much more serious if contracted by an adult. A large outbreak of gastro-enteritis due to contamination of a water supply by rotaviruses was reported in Arizona. The enteric Adenoviruses 40 and 41 are almost as important as Rotavirus as aetiological agents of viral gastro-enteritis in children (Cruz et al., 1990). Although they have not been confirmed as waterborne pathogens they are more stable in water than either Poliovirus 1 or Hepatitis A virus, and are able to survive conventional sewage treatment more effectively than other enteroviruses. They are widespread in water and so it is likely that adenoviruses are indeed waterborne pathogens (Enriques and Gerba, 1995). Astroviruses are another common cause of gastro-enteritis in young children, with 75% of those aged between 5 and 10 years shown to have the astrovirus antibody in the UK (Kurtz and Lee, 1978), with serotype 1 shown to be the major infection-causing Astrovirus (Lee and Kurtz, 1994). As with bacterial infections, many incidents of viral disease associated with drinking water have been attributable to untreated or inadequately treated water or to defects within the distribution system (Craun, 1988). Gerba and Rose (1990) have produced an excellent review on viruses in source and drinking waters.

Outbreaks of waterborne viral disease, other than infectious hepatitis, are difficult to recognize because viruses tend to cause non-apparent or latent infections (Tyler, 1985). Each year a large percentage of reported cases of waterborne disease are of unknown aetiology (Galbraith, Barnett and Stanwell-Smith, 1987; Herwaldt et al., 1992). One possible explanation for such defects in the data is that epidemiological methods are not adequate enough to detect low level transmission of viral diseases via water. This is because a single viral type may produce a wide variety of symptoms which may not be attributable to a single aetiologic agent; also different viruses can produce similar symptoms (Tyler, 1985). Therefore it can be almost impossible to establish the original cause of the outbreak.

Viruses are usually excreted in numbers several orders of magnitude lower than those of coliforms (APHA, 1992). Because they only multiply within living susceptible cells, their numbers cannot increase once excreted. Once in a cell-free state, their survival and infectivity in the aquatic environment depends on a variety of biotic (i.e. type of virus, bacterial and algal activity and predation by protozoa) and abiotic factors (i.e. temperature, suspended matter, pH, salinity, ultra-violet light penetration, organic compounds, adsorption to suspended matter and aggregation) (Geldenhuys and Pretorius, 1989). Temperature is considered to be the most important factor influencing viral destruction outside the host cell. They are rapidly inactivated once exposed to temperatures in excess of 50 °C (Bitton, 1978). Suspended solids provide a certain degree of protection for viruses. Adsorption on to organic matter can prevent inactivation by ultra-violet light. Once adsorbed the viruses can settle from suspension and survive for long periods in sediments to become resuspended if the water becomes turbulent (Watkins and Cameron, 1991).

Sewage treatment, virus dilution, natural inactivation, water treatment and other factors combine to reduce viral numbers to a few survivors in large volumes of water (Metcalf, 1978). In reality viruses generally pass unaffected through wastewater treatment plants and so will be found in surface waters receiving both treated and untreated sewage (Gray, 1992). In the more developed regions of the world, the possibility of viral transmission of waterborne disease depends on the ability of minimum quantities of virus to cause infections (APHA, 1992). The minimum infectious dose for many infectious viruses is unknown but is thought to be as low as 1–10 infectious units (Watkins and Cameron, 1991). It is critical therefore, that small numbers of virus be detected from relatively large volumes of water. In addition, viral detection methods should satisfy the following criteria:

- They should have a high virus concentration factor (1000–10 000) with a high efficiency of recovery for very low concentrations of viruses.
- They should be able to concentrate and detect all enteric virus types with equally high efficiency.
- The method should be relatively inexpensive (Metcalf, 1972).

There are a number of methods available for the detection of viruses, however, none satisfy all of the above criteria. Even the most recent methods are continually being modified and changed. None of the available techniques have been adequately tested with all viral types of importance to public health. Recovery rates may vary depending on water quality. In addition, some methods require large pieces of equipment for sample processing. Identification procedures require cell cultures and related laboratory facilities – such requirements are generally beyond the financial and technical capabilities of most laboratories (APHA, 1992). Detecting viruses in water through the recovery of infectious virus involves three steps:

1. collecting of representative samples;
2. concentrating the viruses in the sample;

3. identifying and estimating quantities of the concentrated virus (APHA, 1992).

Viral numbers are generally so low that their detection is virtually impossible unless they are first concentrated. Many of the concentration techniques available are based on one of two principles: either the filter absorption–elution systems, or ultra filtration systems (Sobsey, 1975).

Filter absorption–elution systems are based on the fact that viruses present in water containing little or no organic matter will absorb to the surfaces of microporous cellulose ester or fibreglass filters under conditions of low pH or in the presence of polyvalent cation salts or both (Sobsey, 1975). Filters may be electropositive or electronegative. Absorbed viruses are then eluted from the filter surface by pressure-filtering a small volume of eluent fluid through the filter *in situ* (APHA, 1992). Despite certain limitations, this technique is considered to be the most promising and is the most widely used technique for virus concentration. The basis for virus concentration by ultra filtration is that water and microsolutes can be driven through a microporous membrane by pressure while microsolutes such as viruses and other high molecular weight materials are retained and thereby concentrated (Sobsey, 1975). The ultrafiltration process described in Standard Methods involves placing the sample in a cellulose dialysis bag and exposing it to polyethylene glycol (PEG), a hygroscopic material. The viruses retained in the dialysis bag are recovered by opening the bag, collecting the remaining sample and eluting any viruses which may have been adsorbed onto the inner walls of the bag. The collected concentrate and eluate are combined and assayed for viruses (APHA, 1992).

After removal of any sample toxicity, viruses may be isolated using tissue culture assays (West, 1991). At present there is no single host system which can be used for all enteric viruses and some, notably the Hepatitis A virus, the human Rotavirus and Norwalk-like viruses, cannot be assayed using routine tissue cultures (APHA, 1992). Advances in immunochemistry, tissue culture and genetic engineering have seen the development of more rapid techniques for the identification of viruses. Such techniques include immunofluorescence, radio-immunoassay (RIA) and enzyme-linked-immuno-sorbent-assay (Table 6.4).

As with bacteria, the development of gene probes for specific viruses has introduced the potential for fast and sensitive assays. Specific probes already exist for rotaviruses, enteroviruses and adenoviruses (Watkins and Cameron, 1991). Gene probes have been found to be as sensitive as, if not more sensitive than, tissue culture techniques. They have the added advantage of not relying on tissue culture and therefore

Table 6.4 Rapid techniques for enteric virus detection (Gerba, Margolin and Hewlett, 1989)

Technique	Assay time (hrs)	Detection limit (pfu)	Max. vol. which can be assayed (mL)	Comment
Immuno-electron microscopy (IEM)	2–4	10^5–10^6	0.05	Requires specific antibody against each virus type
Direct immuno-fluorescence (IF)	2–4	10^5–10^6	0.1	"
IF of infected cell culture	24–48	1	0.5–1.0	"
Enyzme linked immuno-sorbent assay (ELISA)	2–4	10^3–10^5	0.2	"
Radio-immuno-assay (RIA)	2–4	10^4–10^5	0.2	"
Gene probe	2–48	1	30–100	One probe can detect related viruses

offer a cheaper alternative. At present the main limitations associated with using gene probes for virus detection is that there is no distinction made between viable and non-viable cells. Also, they cannot determine the infectivity of the viruses. However, it is important to remember that the use of gene probes for viruses is relatively recent and initial problems are to be expected.

For many years chlorination was considered to be effective in preventing contamination of water supplies by viruses. This conclusion was drawn from the results of epidemiological studies, where it has been repeatedly shown that outbreaks due to viral contamination occur largely in situations where there is inadequate or no chlorination (Galbraith, Barnett and Stanwell-Smith, 1987; Herwaldt et al., 1992). However, in recent years there has been a noticeable change in this situation. Enteric viruses have been isolated from drinking waters which have been treated by chlorination or other processes such as ozonation and chemical coagulation. Such drinking waters contain chlorine levels originally thought to be virucidal (0.1–0.3 mg/L) (West, 1991). Initially the ability to survive chlorination was thought to be due to a lack of contact time with chlorine (Melnick and Gerba, 1980). It is now well established that some enteric viruses are more resistant to chlorination than coliforms (Shaffer, Metcalf and Sproul, 1980). Peterson et al. (1983) showed that exposure to over 2 mg/L of chlorine for 30 minutes was required to inactivate the infectivity of the Hepatitis A virus, while exposure to as much as 5–6 mg/L chlorine for 30 minutes may be required to destroy the infectivity of the Norwalk virus (Keswick *et al.*, 1985). It has been suggested that such resistance may be due to the protective effect of viral aggregation (West, 1991).

Enteroviruses have been recovered from waters which are free from indicator organisms. There has been considerable debate over whether or not there is continual low-level viral contamination of drinking water which subsequently results in sporadic viral infections among consumers (Sellwood and Dadswell, 1991). Such outbreaks largely go undetected due to the large proportion of asymptomatic infections and varied symptomatology in those individuals experiencing such infections (Feacham *et al.*, 1983). This view, while popular in the USA has not afforded much attention in Europe where it is considered that there is no evidence to substantiate the existence of low-level transmission (Feacham *et al.*, 1983; Sellwood and Dadswell, 1991). The possibility of low-level transmission has considerable significance in terms of determining the level of viruses allowable in drinking waters and more especially the costs involved in implementing stricter viral standards, particularly in developing countries where there is a severe shortage of financial and technical resources (Feacham *et al.*, 1983).

6.4 BACTERIA

In addition to viruses and protozoan parasites, the last 20 years have also seen the emergence of new waterborne diseases specifically associated with bacteria. These opportunistic bacteria are usually found as part of the normal heterotrophic bacterial flora of aquatic systems (Reasoner, 1992) and may also exist as part of the normal body microflora (Table 6.5). Of course the traditional waterborne bacterial pathogens of temperate regions, such as *Salmonella* (typhoid, paratyphoid), *Shigella* (bacterial dysentery), and *Vibrio cholerae* (cholera) remain serious threats to water quality. These are considered briefly at the end of this section.

OPPORTUNISTIC BACTERIAL PATHOGENS

These organisms are normally not a threat to healthy individuals, however, under certain circumstances they can lead to infection in certain segments of the community, particularly new-born babies, the elderly and the immuno-compromised (Bitton, 1994). It is thought that numerous hospital acquired infections are attributable to such organisms (De Zuane, 1990). Payment, Franco and Siemiatycki (1993)

Table 6.5 Opportunistic bacterial pathogens isolated from drinking water (Reasoner, 1992)

Acinetobacter spp.
Achromobacter xylosoxidans
Aeromonas hydrophila
Bacillus spp.
Campylobacter spp.[*]
Citrobacter spp.
Enterobacter aerogenes
E. agglomerans
E. cloacae
Flavobacterium meningusepticum
Hafnia alvei
Klebsiella pneumoniae[*]
Legionella pneumopila[*]
Moraxella spp.
Mycobacterium spp.
Pseudomonas spp. (non-aeruginosa)[*]
Serratia fonticola
S. liquefaciens
S. marcescens
Staphylococcus spp.[*]
Vibrio fluvialis[*]

[*]Indicates that the organism may be a primary pathogen.

carried out an eighteen-month epidemiological study of gastro-intestinal illness on a number of families. The drinking water consumed met current bacteriological and physico-chemical quality standards, but a significant level of gastro-intestinal illness was reported. A weak association between the level of illness and heterotrophic bacterial numbers was observed. Further analysis revealed that bacteria growing at 35 °C were responsible for the observed effects. Observations such as these suggest that bacteria generally considered harmless may in fact be disease causing, which raises concerns about the safety of bacterial growth within the distribution system, particularly as HPC in excess of 500 cfu/mL tend to mask coliform occurrences (section 5.7). It is difficult to assess the health implications of these organisms for a number of reasons:

- the lack of data on the occurrence and densities of these organisms in water;
- the lack of data on infectious doses required to establish infection;
- the lack of data on the incidence of human disease caused by water-borne exposure to such organisms;
- the interactive effects of exposure to mixed types and densities of these organisms;
- the range of susceptible individuals in the exposed population;

- the effectiveness of treatment procedures and post-disinfection for control of these agents;
- the need for good detection methodologies that would allow adequate surveillance and monitoring for such organisms.

Some of the organisms listed in Table 6.5 are also to be considered as primary pathogens, meaning that they are also capable of being primary disease causing agents rather than secondary invaders. Of these organisms, those of particular concern at present include *Campylobacter* spp., enterotoxigenic *E. Coli*, *Mycobacteria* spp. and aeromonads.

Campylobacter spp.

Campylobacter are Gram-negative spirally shaped bacteria 2000–5000 nm in length comprising 2–6 coils. They have a single polar flagellum giving them a characteristic darting motility. The bacteria are oxidase-positive and reduce nitrates but are unable to produce acid in the presence of carbohydrates. Although discovered in the late nineteenth century they were not isolated from diarrhoetic stool specimens until 1972. It is only since the development of a highly selective solid growth medium, allowing culture of the bacterium, in 1977 that its nature has been revealed (Bolton *et al.*, 1982). *Campylobacter* species have been isolated from both fresh and estuarine waters with counts ranging from 10–230 campylobacters per 100 mL in rivers in North-west England. Although the epidemiology of human campylobacter infections is yet to be fully elucidated, certain sources of infection are well established. There were 27 000 reported outbreaks of campylobacter enteritis in the UK during 1987, rising to over 30 000 in 1990, causing severe acute diarrhoea, and it is now thought that *Campylobacter* is the major cause of gastro-enteritis in Europe, the USA and other parts of the world, being more common than *Salmonella* (Andersson and Stenstrom, 1986; Blaser *et al.*, 1986). In the USA, the annual incidence of this organism is 30–60 per 100 000 of the population (Skirrow and Blaser, 1992). In developing countries, outbreaks of campylobacter enteritis are a major cause of morbidity and mortality in the first two years of life. While campylobacter enteritis is essentially a food-borne disease, with the most important reservoirs of the bacterium being meat, in particular poultry, and unpasteurized milk, waterborne transmission has been implicated in several large outbreaks (Mentzing, 1981; Taylor *et al.*, 1983; Vogt *et al.*, 1982). Waterborne transmission of *Campylobacter* occurs in untreated, contaminated waters, in situations where faulty disinfection has occurred or where waters have been contaminated by birds and animals (Tauxe, 1992). For example, 3000 of a total population of 10 000 developed campylobacter enteritis from an inadequately chlorinated mains supply in Bennington, Vermont (Vogt et al., 1982), while 2000 people who drank unchlorinated mains water contaminated with faecally polluted river water in Sweden also contracted the disease (Mentzing, 1981). Water is either contaminated directly by sewage which is rich in *Campylobacter* (Marcola, Watkins and Riley, 1981) or indirectly from animal faeces. Household pets, farm animals and birds are all known to be carriers of the disease (Fox, Zanotti and Jordon, 1981; Fox, Ackerman and Newcomer, 1983; Hill and Grimes, 1984; Sticht-Groh, 1982). There is a definite seasonal variation in numbers of campylobacters in river water, with greatest numbers occurring in the autumn and winter. This is opposite to the seasonal variation of infection in the community with number of infections rising dramatically during May and June (Jones and Telford, 1991).

The genus comprises two distinct groups, one that grows at 42 °C and is considered thermophilic, which includes both the important human pathogens *C. jejuni* and *C. coli*, and the other which grows only at 25 °C but not at 42 °C. Serotyping of isolates revealed that *C. jejuni* serotypes common in human infections were especially common downstream of sewage effluent sites, confirming sewage effluents as important sources of *C. jejuni* in the aquatic environment. Gulls are known carriers and can contaminate water supply reservoirs while they roost. Dog faeces, in particular, are rich in the bacterium (Svedhem and Norkans, 1980). In a UK study *C. jejuni* was isolated from 4.6% of 260 specimens of dog faeces sampled, while *Salmonella* spp. were isolated from only 1.2% (Wright, E.P., 1982). Other studies have shown the incidence of *C. jejuni* amongst dogs to range from 7% to 49% (Bruce, Zochowski and Fleming, 1980; Holt, 1980). So dog faeces can cause contamination of surface waters during storms as surface run-off removes contaminated material from paved areas and roads. An outbreak affecting 50% of a rural community based in northern Norway was traced to contaminated faecal deposits from sheep grazing the banks of a small lake, which were washed into the water during a heavy storm that melted the snow on the banks. The water supply for the village came directly from the lake without chlorination (Gondrosen et al., 1985). Natural aquatic systems in temperate areas are generally cool and research has shown that campylobacters can remain viable for extended periods in streams and groundwaters. Survival of the bacterium decreases with increasing temperature but at 4 °C survival in excess of 12 months is possible. The incidence of *Campylobacter* in water can be estimated by an MPN technique (Skirrow and Blaser, 1992). Isolation of this organism from properly treated and disinfected waters has not been reported, thus implying that current water treatment practices are adequate for elimination of *Campylobacter* spp. (Blaser, Taylor and Feldman 1982).

Enterohaemorrhagic *E. coli* (*E. coli* 0157:H7)

Most strains of the bacterium *Escherichia coli* are part of the normal microbial flora of the gastro-intestinal tract of warm-blooded animals including human beings. A number of strains of *E. coli* are pathogenic and cause characteristic gastro-enteritis. Pathogenic *E. coli* are classified into four broad groups based on virulence properties, clinical syndrome, epidemiology and 0:H serogrouping. These are listed below with examples of common serotypes causing gastro-intestinal illness given in parentheses.

1. enteropathogenic *E. coli* (18, 26, 44, 86, 111, 114, 119, 125, 126)
2. enteroinvasive *E. coli* (28*ac*, 112*ac*, 136, 143, 144, 152, 164)
3. enterotoxigenic *E. coli* (6, 8, 15, 25, 27, 63, 78, 115, 148, 153, 154)
4. enterohaemorrhagic *E. coli* (157)

Escherichia coli 0157:H7 belongs to group (4), and is an atypical faecal coliform which causes haemorrhagic colitis and haemolytic–uraemic syndrome and is a leading cause of kidney disease in children. In the USA approximately 20 000 cases due to this organism are reported each year (Finelli *et al.*, 1994), while 650 cases were reported in the UK during 1994 (Maule, 1996). Like *Campylobacter jejuni*, this organism is generally associated with food, in particular beef and milk (Hancock *et al.*, 1994; Neill, 1994), but in recent years has been implicated in a number of waterborne disease outbreaks (Dev, Main and Gould, 1991; McGowan, Wickersham and Strockbine, 1989). However, the organism is still restricted in its distribution. For example a survey of 1267 samples of drinking and recreational waters in northern Greece failed to isolate the serotype and it appears not to be present in potential animal reservoirs of the disease (Arvanitidou, Constantinidis and Katsouyannopoulos, 1996; Kansouzidou *et al.*, 1991). The number of organisms required to initiate infection is thought to be <100, and at present there is no specific treatment for the disease, making the emergence of this new strain of *E. coli* particularly worrying (Maule, 1996). In Cabool, Missouri (USA) *E. coli* 0157:H7-contaminated water resulted in 240 confirmed cases of diarrhoea and four deaths (Geldreich *et al.*, 1992). In all these outbreaks, infection was the result of faecally contaminated surface water or arose where inadequate disinfection had occurred.

Detection of *E. coli* 0157:H7 cannot readily be done using standard faecal coliform methodology as the organism does not grow well at 44 °C in non-selective media and will not grow above 41 °C in selective media. However, as it is known not to be able to ferment sorbitol at 37 °C, total coliforms with a sorbitol negative result are presumptive *E. coli* 0157:H7. These organisms can also be differentiated from other *E.*

coli strains using phase typing or pulsed field gel electrophoresis (Reilly, 1995). Final identification is done by confirming for H7 antigens (Arvanitidou, Constantinidis and Katsouyannopoulos, 1996; Reasoner, 1992), although other techniques such as pyrolysis mass spectrometry are also under investigation (Freeman *et al.*, 1995). The revised standard methods for microbial analysis of drinking water (Department of the Environment, 1994a) proposes a tentative method for the isolation of *E. coli* 0157:H7 based on an enrichment in modified peptone water followed by sub-culture to modified sorbitol MacConkey agar and selection of non-sorbitol fermenting colonies. The method was originally developed for screening milk samples and has yet to be fully evaluated for water.

Mycobacteria

Traditionally, mycobacteria were regarded as environmental contaminants or as transient colonizers in humans. However, they are now recognized as being opportunistic pathogens of considerable significance. Table 6.6 lists the species of opportunistic mycobacteria that commonly cause disease in humans, including pulmonary disease and cervical lymphadenopathy as well as localized and soft tissue infections (Jenkins, 1991). Disease associated with these bacteria is steadily rising, particularly among patients with AIDS. Disseminated mycobacterial disease is now the third most common opportunistic terminal infection in patients with AIDS (Du Moulin and Stottmeier, 1986).

It is well established that mycobacteria are commonplace in all types of aquatic environments, including estuaries, ocean water, groundwater, surface waters and distribution systems (Jenkins, 1991). The majority of waterborne mycobacterial outbreaks are attributable to treatment deficiencies such as inadequate or interrupted chlorination, but other factors may also influence the growth of this organism in water supplies, such as pitting and encrustations found inside old water pipes which protect bacteria from exposure to free chlorine (Du Moulin and Stottmeier, 1986). Mycobacteria can also colonize areas where water is moving slowly, as in water distribution systems in large buildings such

Table 6.6 Opportunistic mycobacteria that commonly infect humans (Jenkins, 1991)

Mycobacterium avium	*Mycobacterium malmoeuse*
Mycobacterium chelonae	*Mycobacterium marinum*
Mycobacterium fortuitum	*Mycobacterium scrotulaceum*
Mycobacterium intracellulare	*Mycobacterium szulgai*
Mycobacterium kansaii	*Mycobacterium xenopi*

as blocks of flats, offices and hospitals, thus continuously seeding the system (Du Moulin and Stottmeier, 1986).

Detection and enumeration of mycobacteria from water samples involves the use of membrane filtration and selective and inhibitory media. The major problem in examining these organisms is the extended incubation period of up to 30 days that is required (Reasoner, 1992).

AEROMONADS

Aeromonas spp. have been isolated from a number of water sources, both raw and treated (Burke *et al.*, 1984a,b). This organism has been implicated as the causative agent in a number of waterborne outbreaks and is now recognized as an opportunistic pathogen (Schbert, 1991). Aeromonads are considered to be an important, and often fatal, cause of non-gastrointestinal illness in immuno-compromised individuals (Schbert, 1991).

Aeromonads have been isolated from both chlorinated and unchlorinated drinking water supplies (Burke *et al.*, 1984a,b) occurring in greatest numbers during the summer months. They have also been isolated in waters containing no *E. coli* and few total coliforms (Schbert, 1991) which raises the question as to how adequate the coliform index is for evaluating water quality. Studies by Versteagh *et al.* (1989) have shown that the addition of copper to drinking water considerably reduces the number of aeromonads present in a sample.

In addition to these organisms, which are considered as emerging problems for water quality, Reasoner (1992) has also identified a number of other organisms which may become significant as waterborne pathogens in the future (Table 6.7). Present standards relating to the microbiological quality of water do not take into account the potential pathogenicity of these organisms. While many opportunistic organisms present no threat to healthy individuals, they pose a high health risk to the more vulnerable members of the community, i.e. the elderly, the very young and the immuno-compromised. As these groups within the population are steadily increasing, it is likely that in the future greater emphasis will have to be placed on opportunistic bacterial populations. Reasoner (1992) argues that there is a tendency to look for a single causative agent for a waterborne outbreak. In the future, he concludes, it is likely that 'the collective impact of simultaneous and repeated exposures to variable levels of several opportunistic bacteria will be considered, rather than the effects of an individual organism'.

Perhaps of greatest concern in relation to these organisms is that the coliform count does not reflect their incidence in water. In addition to

Table 6.7 Emerging and potential waterborne opportunistic pathogens (Reasoner, 1992)

Organism	Old/emerging	Potential candidates
Anaerobiospirillum succiniciproducens		X
Aeromonas spp.	X	
Campylobacter spp.	X	
Escherichia coli 0157:H7	X	
Helicobacter pylori		X
Mycobacterium spp.	X	
Plesimonas shigelloides		X
Vibrio fluvialis		X

the public health significance, the presence of these emerging organisms indicates an inadequacy in the barriers in place to protect public health, leading to the exposure of the population to potentially polluted waters.

PRIMARY BACTERIAL PATHOGENS

The various serotypes that make up the genus *Salmonella* are now possibly the most important group of bacteria affecting the health of both humans and animals in Western Europe. For humans this is undoubtedly due to the elimination of other classical bacterial diseases through better sanitation, higher living standards and the widespread availability of antibiotic treatment. *Salmonella* is commonly present in raw waters but only occasionally is isolated from finished waters, chlorination being highly effective at controlling the bacteria. Typical symptoms of salmonellosis are acute gastro-enteritis with diarrhoea, and it is often associated with abdominal cramps, fever, nausea, vomiting, headache and, in severe cases, even collapse and possible death. Compared to farm animals the incidence of salmonellosis in humans is low and shows a distinct seasonal variation. A large number of serotypes are pathogenic to humans and their low frequency of occurrence varies annually from country to country, with wild, domestic and farm animals often acting as reservoirs of human salmonellosis. Low-level contamination of food or water rarely results in the disease developing, because 10^5–10^7 organisms have to be ingested before development. Once infection has taken place then large numbers of the organisms are excreted in the faeces (more than 10^8 per gram). Infection can also result in a symptomless carrier state in which the organism rapidly develops at localized sites of chronic infection, such as the gall bladder or uterus, and is excreted in the faeces or other secretions. Water resources can become contaminated by raw or treated wastewater as well as by effluents from abattoirs and animal processing plants (Gray, 1992).

Characteristically *Salmonella* spp. conform in general to the Enterobacteriaceae but can be further differentiated biochemically into four sub-genera (I to IV). Only species in sub-genus I are known to be human pathogens and are B-galactosidase-negative. The most serious diseases associated with specific species are typhoid fever (*Salmonella typhi*) and paratyphoid (*Salmonella paratyphi* and *Salmonella schottmuelleri*). The last major outbreak of typhoid in Britain occurred in Croydon, Surrey during the autumn of 1937, when 341 cases were reported resulting in over 40 deaths (Galbraith, Barnett and Stanwell-Smith, 1987). There have been five minor outbreaks of typhoid and three of paratyphoid since then. The number of reported cases of typhoid fever in the UK has fallen to less than 200 per annum, 85% of these cases being contracted abroad. Of the remainder, few are the result of drinking contaminated water. Although *Salmonella paratyphi* is recorded in surface waters all over the British Isles, there have been less than 100 reported cases of paratyphoid fever reported annually. Typhoid has also been largely eliminated from the USA, although in 1973 there was an outbreak in Dade County in which 225 people contracted the disease from contaminated well water (Craun, 1986). However, typhoid fever is still common in countries where there is neither a safe water supply nor adequate sewage treatment (Hornick, 1985; Ohasi, 1988). Salmonellosis carries a significant mortality amongst those with acquired immuno-deficiency syndrome (AIDS) and poses significant problems in its management (Wong, S.S.Y. *et al.*, 1994).

Shigella causes bacterial dysentery or shigellosis and is one of the most frequently diagnosed causes of diarrhoea in the USA (Blaser, Pollard and Feldman, 1984). Shigellosis is a problem of both developed and developing countries with the Eastern Mediterranean countries considered as an endemic region for the disease (Samonis *et al.*, 1994). The bacteria of the genus are Gram-negative non-motile rods which are oxidase-negative. With the exception of *S. dysenteriae* type 1, the genus is catalase positive. The bacterial genus is rather similar in epidemiology to *Salmonella* except that it rarely infects animals and does not survive quite so well in the environment. When the disease is present as an epidemic it appears to be spread mainly by person-to-person contact, especially between children, shigellosis being a typical institutional disease occurring in over-crowded conditions. However, there has been a significant increase in the number of outbreaks arising from poor quality drinking water contaminated by sewage (Samonis *et al.*, 1994). *Shigella* may also be carried asymptomatically in the intestinal tract. Of the large number of species (over 40), only *S. dysenteriae*, *S. sonnei*, *S. flexneri* and *S. boydii* are able to cause gastro-intestinal disease.

Shigella sonnei and *S. flexneri* account for over 90% of isolates, although it is *S. dysenteriae* type 1 which causes the most severe symptoms due to the production of the shiga toxin. The number of people excreting *Shigella* are estimated as 0.46% of the population in the USA, 0.33% in Britain and 2.4% in Sri Lanka (Dart and Stretton, 1977). In England and Wales notifications of the disease rose to between 30–50 000 per year, falling below 3000 per year in the 1970s. However, in the 1980s notifications doubled to nearly 7000 per year (Galbraith, Barnett and Stanwell-Smith, 1987).

Cholera is thought to have originated in the Far East where it has been endemic in India for many centuries. In the nineteenth century the disease spread throughout Europe where it was eventually eliminated by the development of uncontaminated water supplies, water treatment and better sanitation. It is still endemic in many areas of the world, especially those which do not have adequate sanitation and, in particular, in situations where the water supplies are continuously contaminated by sewage. This is the cause of the most recent epidemic raging through South America which started in Peru and at the beginning of 1992, some 400 000 people had so far contracted the disease. However, over the past ten to fifteen years the incidence and spread of the disease has been causing concern which has been linked to the increasing mobility of travellers and the speed of travel. Healthy symptomless carriers of *Vibrio cholerae* are estimated to range from 1.9 to 9.0% of the population (Pollitzer, 1959). However, this estimate is now thought to be rather low with a haemolytic strain of the disease reported as being present in up to 25% of the population. The holiday exodus of Europeans to the Far East, which has steadily been increasing since the mid 1960s, will have led to an increase in the number of carriers in their home countries and an increased risk of contamination and spread of the disease. There have been nearly 50 reported cases of cholera in the UK between 1970 and 1986, although no known cases have been waterborne (Galbraith, Barnett and Stanwell-Smith, 1987). While the disease in now extremely rare in the developed world, major waterborne outbreaks occur in developing countries, war zones and disaster areas. For example there have been over 500 000 cases of cholera in Peru during the period between 1991 and 1994.

Up to 10^6–10^7 organisms are required to cause the illness, so cholera is not normally spread by person-to-person contact. It is readily transmitted by drinking contaminated water or by eating food handled by a carrier, or which has been washed with contaminated water, and is regularly isolated from surface waters in the UK (Lee *at al.*, 1982). An infected person or symptomless carrier of the disease excretes up to 10^{13} bacteria daily, enough to theoretically infect 10^7 people! It is an intestinal disease with characteristic symptoms, that is, sudden diarrhoea with copious watery faeces, vomiting, suppression of urine, rapid dehydration, lowered temperature and blood pressure and complete collapse. Without therapy the disease has a 60% mortality rate, the patient dying within a few hours of first showing the symptoms, although with suitable treatment the mortality rate can be reduced to less than one percent.

The bacteria of the genus are Gram-negative curved or comma-shaped rods that are actively motile. They are aerobic, facultatively anaerobic, oxidase-positive and able to grow at pH 8.6. *Vibrio cholerae* has been sub-divided in to over 80 0-serovars, although epidemic cholera is caused by toxin-producing strains of the 01 serovar. Recent evidence shows that epidemic cholera is caused by two different somatic serotypes 01 and 0139 (Bhattacharya *et al.*, 1993). 0139 *V. cholerae* was first reported in Madras in 1992 and appears to be as virulent as the 01 serovar (Swedlow and Ries, 1993). It is now widely distributed and threatens to become the eighth worldwide cholera epidemic (pandemic) (Bodhidatta *et al.*, 1995; Mukhopadhyay *et al.*, 1995). Other species of the genus can also cause diarrhoea although normally less severe than 01 *V. cholerae* (e.g. *V. parahaemolyticus*, *V. fluvialis* and *V. mimicus*). The bacteria are natural inhabitants of brackish and saline waters and are rapidly inactivated under unfavourable conditions such as high acidity or high organic matter content of the water, although in cool unpolluted waters *Vibrio cholerae* will survive for up to two weeks. Survival is even greater in estuarine and coastal waters.

6.5 CONCLUSIONS

Current regulations for water quality require that pathogens should not be present at concentrations which might present a risk to public health. While such regulations are largely upheld with regard to the majority of bacterial pathogens, there is growing evidence that this is not the case with viral and protozoan pathogens. Recent years have seen a rise in the incidence of waterborne outbreaks attributable to these agents in waters considered safe to drink under present coliform standards and this has understandably given rise to concern regarding the validity of such standards for indication of these organisms.

Human enteric viruses and protozoan parasites possess certain traits which aid waterborne transmission and which have contributed to their increase in recent years (West, 1991). These include:

- They can be excreted in faeces in large numbers during illness.
- Conventional sewage treatment has failed to remove them.
- They can survive as an environmentally robust form or they demon-

strate resilience to inactivation whilst in an aquatic environment.

- They are largely resistant to common disinfectants used in drinking water treatment.
- Most importantly, they only require low numbers to elicit infection in hosts consuming or exposed to water.

These factors are compounded by the difficulty of isolating and accurately detecting these pathogens in both treated and untreated effluents as well as in both surface and ground water resources. The identification of several viral agents is problematic because some of these agents cannot be propagated in the laboratory. For both viruses and protozoa, large sample volumes must be examined in order to detect small numbers of organisms. Propagation and identification may take several days. In addition, by the time the outbreak is eventually recognized, it is usually long after the initial contamination event and that water is no longer available for examination. This is particularly the case with protozoan pathogens when a minor operational error during backwashing sand filters may result in the breakthrough of cysts and oocysts into the treated water. There is also the question of determining the viability of such organisms, particularly *Cryptosporidium* and *Giardia*, as present detection methods do not allow for such distinction. Recent developments in PCR and gene probe technology show considerable promise particularly in relation to determining viability. However, as with the detection of bacterial pathogens, the use of PCR is very much hampered by the presence of substances such as humic acids, commonly found in water originating from upland reservoirs and acidic rivers.

An emerging issue for the water industry is the virulence of opportunistic bacterial pathogens which are frequently found as part of the heterotrophic population in distribution and home plumbing systems. The extent of this problem has not yet been fully established, but studies have shown a definite association between heterotrophic bacteria growing at 35 °C and gastro-intestinal illness (Payment *et al.*, 1991). The problem is further complicated by a number of bacteria which have recently been implicated in waterborne outbreaks for the first time.

At present it would seem that the water industry is faced with the problem that however meticulously it adheres to the accepted practice of using indicator organisms it cannot guarantee that the drinking water it supplies will be free from all pathogens. Neither does it appear that adequate risk assessment can be carried out using such indicators.

The future of the coliform index in water analysis

7

7.1 INTRODUCTION

The public health importance of clean drinking water requires an objective test methodology to evaluate the effectiveness of treatment procedures and to establish drinking water standards (Bonde, 1977). The group which has best met the criteria established for such a methodology has been the coliform group. Since its development in the 1890s, the use of this group as an indicator of faecal contamination has served as the principal and often sole criterion of water quality. The success of this approach is borne out by the dramatic decrease in the occurrence of classical waterborne diseases such as cholera and typhoid fever since the turn of the century. In general, the coliform test has proved a practical measurement of treatment effectiveness, although there is much debate concerning the adequacy of the coliform index and its ability to determine the potability of drinking water.

Any criticism of the coliform index must take into account that it was developed nearly a century ago and therefore reflects the disease profile of that time and not of the 1990s. This has tended to mean that monitoring data being generated under current practices are more representative of the historical record of water produced yesterday rather than the quality of water being released for supply at ·the moment (Geldreich, 1992). Waterborne disease is now known to be caused by a much broader spectrum of microbes than just the enteric bacteria. In recent years, there has been a proliferation in the number of waterborne outbreaks attributable to viral and protozoan agents. This has given rise to considerable concern in the water industry, particularly as these organisms are generally more resistant to conventional treatment than bacterial pathogens, more difficult to detect and not associated with the coliform index. The concern is that a negative coliform index no longer guarantees that a water is free from all pathogens. Primary focus to date has always been on microbial risks due to gastro-intestinal pathogens in contaminated waters. While this objective remains of paramount concern for water quality, the appearance of opportunistic pathogens that are normally part of the heterotrophic bacterial flora of aquatic systems is emerging as a serious public health concern (Geldreich, 1996). These organisms are primarily a problem for young children and for those who are immuno-compromised, and because they do not pose any great problem for healthy individuals, their significance as pathogens is often discounted (Reasoner, 1992). Herein lies a significant limitation of coliforms as indicators for, in the future, characterization of a microbially safe water supply must be expanded to include all of the new pathogens whose presence the coliform is apparently unable to indicate.

Water quality guidelines and standards have emerged as the best possible means of balancing the needs of multiple water use. In theory, the setting of standards should be based on sound, logical, scientific grounds, taking into account environmental, social, economic and cultural factors. The question of bacteriological standards for drinking water remains a subject of considerable debate. At present, the coliform is the principal microbial standard mandated in most water quality standards. Many American and European standards adopt a precautionary approach and therefore have tended to be very high. While it is not unreasonable to expect properly treated water to meet such standards, the practicality and costs of achieving them is often not appreciated, nor are the costs of regulation. In Britain, only 20% of water company spending is allocated towards the provision of water supplies. The remaining 80% goes towards meeting EU water quality requirements (Johnstone and. Horan, 1994). In the USA, recent changes in coliform standards have made it significantly more costly and more difficult to supply safe water. It is estimated that as a result of these revisions, the number of US water companies with violations may rise as much as sixfold. Given the uncertainty regarding the use of the coliform as an indicator organism, to base all microbiological water quality standards on its presence or absence in water is somewhat unjustifiable. Much of the demand for high microbiological standards for water quality comes from those who believe that properly treated water should pose a zero risk to the consumer and that anything other than this is unacceptable. Setting such high standards may have the opposite effect in the long run, as the costs of implementation are often so high.

There is a long tradition of developments in legislation and technology being directly transferred from developed to underdeveloped countries without proper consideration for the implications involved. Most water supplies in developing countries are untreated and are often grossly contaminated, sometimes having an indicator bacterial concentration of weak raw sewage (Feacham *et al.*, 1983). Indeed, in many developing countries people are using water of a quality that would not be recommended for recreational activities in Europe and Northern America (McDonald and Kay, 1988). Bacterial waterborne disease remains a serious public health problem in these areas and subsequently there is a serious need for realistic standards based on relevant scientific information. While there may be some doubts about the relevance of the coliform in developed regions, numerous studies have shown that there is little or no justification for its use in tropical environments. In any case, there are many problems in these regions in achieving any microbial water quality standard. Most of the developing countries do not have the resources nor the political will to regulate such standards. The most rational answer is a phased approach based on affordability which will allow developing nations time to develop appropriate standards (Johnstone and Horan, 1994).

There is concern about what appears to be a lack of correlation between coliforms and bacterial as well as viral and protozoan pathogens. However, it is important to remember that the coliform standard is a standard of expedience and is not intended to be a standard of perfection. It embodies the concept of a small but allowable risk of enteric infection and acknowledges that all risk from enteric pathogens cannot be realistically eliminated (Wolf, 1972). In any case, it was never intended that quantitative and qualitative determinations of bacterial indicators would be the sole information to judge health hazards associated with a particular water. There is a certain misconception regarding enteric indicators, in that it is often implied that a knowledge of the levels of these organisms has resulted in the assurance of safety of water supplies. While the use of indicator organisms has contributed enormously to public health microbiology, it is our knowledge of the behaviour of these organisms during water treatment which allows us to use quantitative estimations in water to predict the degree of treatment required to render it safe (Fox, Keller and van Schothoist, 1988). The importance of a sanitary survey cannot be overstated, that is the investigation of possible sources and routes of pollution. Quantitative estimations of contamination can then be made and interpreted according to the particular circumstances prevailing in the system.

Research priorities are now moving from investigation of bacterial pathogens to the evaluation of risks posed by viral and parasitic pathogens in the water supply. The concept of an acceptable risk lies at the heart of many standards though it has yet to be established what constitutes an acceptable risk. Ultimately a cost–benefit approach will have to be used as standards are continually being improved (Fawell and Miller, 1992). Accordingly, it will be necessary to improve education and understanding of the general concept of risk assessment especially among those who become over sensitive to a particular possibility of risk without a proper appreciation of the likely magnitude of that risk (West, 1991).

The search for the ideal indicator system has been going on for many years and although many have been proposed (Chapter 5), none have been found to be perfect. Some have shown a less than absolute correlation with waterborne pathogens, others are too resistant to conventional treatment, too ubiquitous in nature or require complex isolation methodologies unsuitable for routine use. In any case, it is not feasible to have a single indicator for all situations and places and therefore universal pollution indices should be interpreted with caution. Thus, no new indicator system has yet emerged as a successful alternative to the coliform.

At present, the inability to detect indicators or pathogens within a few hours of sample processing is a major limitation in water quality assessment. To overcome this problem, there has been considerable research into developing rapid detection methods. Extremely sensitive and specific techniques are now available for the detection of pathogens. Nonetheless, their usefulness in analysing environmental samples is somewhat limited when the expertise and high costs some of these methods require are taken into account. This is particularly so for smaller water utilities and for developing countries. Future developments in PCR and gene probe technology may eventually mean that the detection of indicator organisms will be replaced by the direct detection of pathogens. There is a certain hesitancy among those involved in water analysis to depart from traditional methods because of their long tradition of use and the considerable data base that exists in relation to these methods. At present, the best solution would appear to be using a combination of traditional and molecular techniques so as to achieve the level of sensitivity and reliability required (Alvarez, Hernandez-Delgado and Toranzos, 1993).

Generations of microbiologists have used the multiple tube technique, carefully ensuring the inversion of the Durham tube without any gas bubble, surely one of the most difficult tasks for the student. The advent of the membrane filtration technique effectively opened up the index to non-microbiologists and the availability of commercially prepared agar plates now means that tests are routinely carried out in laboratories without even the most basic of microbiological equipment. The concept of sterility in sample collection, sample dilution, filtration and plate preparation is not widely appreciated by or adhered to by non-specialists. Over the years the coliform index has changed very little, and it has become one of the classic water quality standard methods, used in laboratories throughout the world. This widespread acceptance of the index may have led to complacency and misuse. Rather like the other traditional water quality tests it has become increasing obvious that the coliform index is no longer adequate and is often being used to perform tasks, or used as the basis of operational decisions, for which it is inappropriate.

A major problem appears to be that few academic microbiologists appear to be interested in water microbiology. The current status of environmental, and in particular water, microbiology in universities is generally poor, although there are notable exceptions. Few graduate microbiologists have a clear understanding of the current issues and problems in drinking water microbiology, and leave university with the same techniques and ideas that were being taught often generations ago. With very little research in the area and the coliform index now being carried out largely by non-microbiologists, the development and introduction of new techniques and indicators for waterborne pathogens will depend on those few dedicated microbiologists largely based in State Laboratories and surprisingly in Civil Engineering Departments around the world.

The coliform index has become such an accepted method that few question the usefulness of the results obtained. Quality control and harmonization of techniques or laboratories is not widespread. While the coliform index is adequate to protect the consumer from the primary bacterial pathogens, it is wholly inadequate to protect them from the new viral and protozoan pathogens that are now responsible for the bulk of waterborne disease in the developed world. No alternative methods have been introduced to ensure drinking waters are tested and consumers adequately protected. With billions of pounds being spent by the new water companies in England and Wales alone to improve drinking water quality, they are using methods to test the safety of water supplies which are over eighty years old.

7.2 THE COLIFORM INDEX TODAY

The problems facing the water industry in relation to microbial safety of drinking water has reached a critical period. The issues that need to be examined include:

LACK OF UNDERSTANDING

Generations of microbiologists and engineers have been taught that direct testing for pathogens is time consuming, expensive, impracticable and unnecessary, and that indicators do the job of water evaluation more effectively. However, there is clearly a lack of understanding at all levels as to what exactly the coliform index does measure and what it is capable of telling us about microbial quality of water overall. The results are only indicative, never conclusive. Even when the coliform tests are negative, we still cannot be sure that the water is microbially safe to drink. For example, nearly every reported outbreak of crypto-sporidiosis has occurred where the water has tested negative for both total and faecal coliforms (section 6.2).

POOR QUALITY CONTROL

The training given in water microbiological testing to both microbiologists and non-microbiologists is often very poor. Sample collection is often carried out by non-competent technical staff so that there is a high risk of contamination. Laboratories are often poorly equipped and staffed, and there is little quality control or harmonization of laboratories so that little is known about the repeatability or reproducibility of the index when used routinely. The coliform index is time-consuming, laborious and a particularly monotonous test method to perform routinely. This, and the very familiarity of the index, has resulted in poor quality control so that results are often suspect.

There is currently a serious lack of investment in research and development in microbial water quality assessment. Investment for the development of new techniques is urgently required. In the interim better training is necessary, especially at universities; there needs to be wider use made of microbiologists to carry out such tests, and finally improved laboratory facilities. But overall better understanding is necessary by those in the water industry and who regulate water quality of the dangers of relying so heavily upon the coliform index as the principal safeguard of water quality from pathogens.

DEFINITION OF WHAT INFORMATION IS REQUIRED

The consumer and water undertaker need to know that the water supplied is free from all pathogens that may cause illness. Potential pathogens include bacteria, viruses, protozoan cysts and oocysts, as well as other micro-organisms. Or at least there is a need to know that the risk to consumers from the presence of these pathogens in water is acceptable. This risk is different for various groups within the community with the young, old and immuno-compromised particularly vulnerable. The problem is how to assess this risk and translate it into realistic standards.

EFFECTIVENESS OF THE COLIFORM INDEX TO SUPPLY THE REQUIRED INFORMATION

From the increasing prevalence of waterborne outbreaks in recent years, it is increasingly clear that the coliform index does not offer an adequate barrier of protection for the community as a whole. This inadequacy is largely offset by excellent medical facilities and rapid recognition of infection. For those who are within the high risk categories, the medical safety net may not be sufficient to avoid unacceptably high mortalities in the event of water contamination occurring.

IS THE COLIFORM INDEX A GOOD ENOUGH SAFEGUARD?

In the developing world, the coliform index must be considered as being wholly inadequate and all water supplies should be considered suspect. Point-of-use treatment or sterilization of water is vital. In developed regions the question can only be answered after consideration of the particular circumstances, although in the broader context it would appear that the answer must also be no. In fact it is becoming increasingly clear that the use of the coliform index on its own is having a negative effect on microbial safety, with water undertakers and consumers accepting a negative faecal coliform result as ensuring microbially safe water. There is an ongoing deterioration in microbial water quality caused not by any negligence on behalf of the water undertaker, but by the general evolution and distribution of pathogens (Gray, 1994). However, supplying water to consumers which could be contaminated either at source or during distribution, and failing to use monitoring techniques adequate to detect such pathogens, could be considered negligent. In the UK the Water Industry Act 1991 requires water supply undertakers to supply consumers with water that is wholesome at the time of supply. Under the Act (section 65) the term wholesome is defined with reference to the Water Supply (Water Quality) Regulations 1989. Here drinking water is regarded as wholesome if:

- It meets the standards prescribed in the regulations.
- The hardness or alkalinity of water which has been softened or desalinated is not below the prescribed standards.
- It does not contain any element, organism or substance, whether alone or in combination, at a concentration or value which would be detrimental to public health.

As protozoans and viruses are the major causes of waterborne outbreaks in the UK, then failure to specifically monitor drinking water for such pathogens seems unacceptable in the light of the legal definition of wholesomeness.

7.3   FUTURE ACTION

Within the current available technology what can be done now to improve microbial water quality testing? It is generally agreed in the light of the new definition of *E. coli* that the use of faecal coliforms should be discontinued and that *E. coli* only should be used as a specific indicator of faecal contamination (Chapter 2). Rapid methods for isolating *E. coli* are needed, with those based on MUG preferred at present. The simple membrane filtration method proposed by Berg and Fiskdal (1988) incorporates MUG in a modified M-T7 agar medium giving a detection level as low as one faecal coliform per 100 ml in just 6 hours. The Colilert test is currently the best test method of its type that is widely available (Chapter 5). More attention must be paid to the enumeration of injured and stressed coliforms and *E. coli* in particular (McFeters, 1990). Fully automated microbiological techniques are already at the development stage. Total coliforms and HPC should only be used to assess raw water quality, treatment efficiency and changes in the quality of finished waters in distribution systems. These tests cannot assess the risk of a waterborne outbreak within the community, but they are useful for routine monitoring to indicate significant reductions in raw water quality that may require additional treatment, failure or inadequacies at the treatment stage, and finally any problems of regrowth in the water mains. It is possible to identify those distribution systems where potential biofilm development can be expected with reasonable accuracy, and so will require intensive surveillance programmes. Clearly the improvement of microbial water quality from treatment plants, and within the distribution system, through better surveillance will significantly reduce the risk of disease. No indicator organism has currently been identified which is able to indicate with any degree of certainty the presence of viral and/or protozoan pathogens. Primary pathogens appear controllable using the dual barrier approach of good water treatment and monitoring of *E. coli*. The lack of a general indicator or a rapid method to detect the presence of all potential pathogens remains the critical dilemma for the water microbiologist. Clearly better resource management and effective wastewater and water treatment plant operation are all required to reduce contamination of supplies through the identification and control of possible sources of infection, and to ensure the optimal removal and destruction of pathogens. However, some form of routine surveillance is required. Risk management may be able to rationalize such surveillance programmes in terms of identifying those resources and supplies at risk, seasonality of infection, and the prediction of infection. Standard isolation methods for viruses are time-consuming and expensive, although bacteriophages offer some potential as an indicator. Gene probe technology is currently the only hope for rapid pathogen-specific surveillance, although the costs will be high initially (Richardson, Stuart and Wolfe, 1991). The current research priority is the development of multiplex probes to detect a wide range of pathogens simultaneously with the sensitivity to identify single organisms within 1-litre samples in under an hour. New standards are required. The proposed EU standards are ambiguous, stating that waters should be pathogen free and yet failing to set specific parameter values for many pathogens. With new standards would come impetus and financial support for much-needed investment in research.

Risk–benefit analysis is urgently required to identify and evaluate the real risks that consumers are now facing. Public warnings must be given whenever there is a doubt pertaining to water quality due to treatment failure or known contamination. Warnings arising from microbial analysis should be based solely on the presence of E. coli or other known pathogens in the water supplies. It must be recognized that immuno-compromised people in particular are at a very high risk level from poor quality water, including the secondary bacterial pathogens (Table 5.6). Therefore an extra barrier (i.e. point-of-use treatment) should always be recommended.

7.4   THE FINAL ANALYSIS

The use of the coliform index as it currently stands (i.e. total and faecal coliform testing) cannot protect consumers from all potential water-borne pathogens in their drinking water. Better microbial assessment methods are required. Legally water undertakers are required to provide water free from all pathogens. Yet we have seen that the concept of zero risk is probably not achievable. So where does the consumer stand? They certainly feel that any risk is unacceptable in terms of the water they drink. So consumers may be forced to take an active role in reducing their personal risk of infection from waterborne pathogens. For those living in high risk areas the only current options are (1) point-of-use devices (i.e. $1\,\mu m$ filtration followed by ultra-violet sterilization) or (2) to drink only bottled water (Gray, 1994). Boiling water may not be sufficient to destroy all pathogens, unless done for at least 10 minutes. It has been suggested that it may be cheaper and more effective for water companies to supply point-of-use treatment systems to consumers in high risk areas rather than try to supply mains water that is guaranteed pathogen-free. Certain water companies in England have been providing bottled water in high nitrate areas for pregnant women and children under three years of age (Gray, 1994). Microbial contamination is by far the most common cause of illness and disease when compared to chemical and physical contaminants. It affects large numbers within the population, often leaving them with serious secondary problems, whereas chemical contamination may only affect those specifically at risk. In real terms the amount of money being spent on reducing nitrate and aluminium in drinking water for example is enormous, yet there is very little clinical evidence to support the perceived risk by consumers to their health. Indeed nitrate-induced infantile methaemoglobinaemia has been shown to be primarily a problem in microbially contaminated waters only, although the numbers of reported cases is extremely low. There have been no cases of water-induced infantile methaemoglobinaemia reported in the UK since 1972 (Gray, 1994). This is clearly not the case with microbial contaminants, and the battle against these pathogens must be constantly reviewed. Therefore in order to reduce the real risk of waterborne outbreaks, as well as to alleviate the perceived risk to consumers reflected in record sales of bottled water and point-of-use treatment systems, more investment is required in tackling the problems of microbial contamination in our water supplies. This of course means investment in better resource management and treatment technology. However, the key to public safety and health will always be effective microbial surveillance of all supplies, both public and private. This requires a concerted effort by government, water undertakers and microbiologists alike to develop new techniques. It also requires us to seek new goals in microbial water quality based on the presence of actual pathogens.

Who is to blame for our current situation? The problem has partially arisen from the change in public attitudes. In the past the main priority was the protection of public health, indeed this was the original motivation for the development of our water and wastewater treatment systems. The concern for our environment has in fact pushed the issue of public health to one side, with water company investment reflecting this. To many in the industry the battle against the classic waterborne diseases of the eighteenth century has been won and to a certain extent this is true. The water industry, however, ignores the emergence of the new pathogens and the inability to detect such pathogens in drinking water being supplied to consumers at their peril.

Finally, is there still a place for the coliform index in water quality assessment? Detection of faecal contamination is still a main priority for water quality control and therefore the presence of any coliform organism in treated drinking water suggests either inadequate treatment or introduction of undesirable material after treatment. While the coliform can still be considered an adequate indicator of treatment efficiency, the same considerations cannot be extended to its capabilities to evaluate the presence of pathogens in drinking water. It is recognized that the coliform index is not a standard of perfection but no matter how meticulously the coliform index is carried out it can not guarantee that drinking water will be free from all pathogens. Neither is the index an adequate criterion for meaningful risk assessment.

There have been enormous advances in the development of alternative indicators and detection methods, and these should be used to their full potential. Unwillingness to change hampers research and the possible development of more meaningful indicators of faecal pollution and of potential health hazards. In conclusion, it would appear that there is still a place for the coliform as a measure of treatment effectiveness, however, based on information now available its presence in water should no longer remain the sole criterion on which the microbiological quality of water is based.

References

Abbazadegan, M., Gerba, C.P. and Rose, J.B. (1991) Detection of Giardia cysts with a cDNA probe and applications to water samples. *Applied and Environmental Microbiology*, 57, 927–31.

Abbott, M.A., Poisz, B.J., Byrne, B.C. *et al.* (1988) Enzymatic gene amplifiaction: qualitative and quantitative methods for detecting proviral DNA amplified in-vitro. *Journal of Infectious Diseases*, 158, 1158–69.

Ainsworth, R.G. (1990) Water treatment for health hazards. *Journal of the Institution of Water and Environmental Management*, 4, 489–93.

Akin, E.W. and Jakubowski, P. (1986) Drinking water transmission of Giardiasis in the US. *Water Science and Technology*, 18, 219–26.

Alexander, L.M. and Morris, R. (1991) PCR and environmental monitoring – the way forward. *Water Science and Technology*, 24, 291–4.

Allsop, K. and Stickler, D.J. (1984) The enumeration of *bacteroides fragilis* group of organisms from sewage and natural waters. *Journal of Applied Bacteriology*, 56, 15–24.

Alvarez, A.J., Hernandez-Delgado, E.A. and Toranzos, G.A. (1993) Advantages and disadvantages of traditional and molecular techniques applied to the detection of pathogens in water. *Water Science and Technology*, 24, 253–7.

Andersson, Y. and Stenstrom, T.A. (1986) Waterborne outbreaks in Sweden causes and etiology. *Water Science and Technology*, 18, 185–90.

APHA (1976) *Standard Methods for the Examination of Water and Wastewater*, 15th edn, American Public Health Association, Washington, DC.

APHA (1992) *Standard Methods for the Examination of Water and Wastewater*, 18th edn, American Public Health Association, Washington DC.

Apte, S.C., Davies, C.M. and Peterson, S.M. (1995) Rapid detection of faecal coliforms in sewage using a colorimetric assay of ß-D-galactosidase. *Water Research*, 29, 1803–6.

Arvanitidou, M., Constantinidis, T.C. and Katsouyannopoulos, V. (1996) Searching for *Escherichia coli* 0157 in drinking and recreational waters in Northern Greece. *Water Research*, 30, 493–4.

Asbolt, N.J., Dorsch, M. and Banens, B. (1995) Blooming *E. coli*. What do they mean?, in *Coliforms and E. coli: Problem or Solution*, abstract of papers and posters of conference held at University of Leeds, September, 1995.

Atlas, R.M. and Bartha, R. (eds) (1992) *Microbial Ecology, Fundamentals and Applications*, the Benjamin/Cummings Publishing Company Inc., Redwood City, CA.

Audicana, A., Perales, I. and Borrego, J.J. (1995) Modification of Kanamycin–Esculin–Azide Agar to improve selectivity in the enumeration of fecal streptococci from water samples. *Applied and Environmental Microbiology*, 61, 4178–83.

Avoort, H.G.A.M. van der, Reimerink, J.H.J., Ras, A. *et al.* (1995) Isolation of epidemic poliovirus from sewage during the 1992–3 type 3 outbreak in the Netherlands. *Epidemiology and Infection*, 114, 481–91.

Badenoch, J. (1990) Cryptosporidiosis – a waterborne hazard (opinion) *Letters in Applied Microbiology*, 11, 269–70.

Barer, M.R. and Wright A.E. (1990) *Cryptosporidium* and water: a review. *Letters in Applied Microbiology*, 11, 271–7.

Barnes, R., Curry, J.I., Elliot, L.M. *et al.* (1989) Evaluation of the 7 hour membrane filter test for quantification of faecal coliforms in water. *Applied and Environmental Microbiology*, 55, 1504–6.

Bascomb, S. (1987) Enzyme tests in bacterial identification. *Methods in Microbiology*, 19, 105–60.

Bayley, S.T. and Seidler, R.J. (1977) Significance of faecal coliform positive *Klebsiella*. *Applied and Environmental Microbiology*, 33, 1141–8.

Belieff, B. and Mary, J.Y. (1993) The 'Most Probable Number' estimate and its confidence limits. *Water Research*, 27, 799–807.

Bej, A.K., Steffan, R.J., Dicesane, J.L. *et al.* (1990) Detection of coliform bacteria in water by using PCR and gene probes. *Applied and Environmental Microbiology*, 56, 307–14.

Bej, A.K., Dicesane, J.L., Haff, L. and Atlas, R.M. (1991) Detection of *E. coli* and *Shigella* species in water by using the polymerase chain reaction and gene probes. *Applied and Environmental Microbiology*, 57, 1013–17.

Benton, C., Khan, F., Monaghan, P. *et al.* (1983) The contamination of a major water supply by gulls (*Larus* sp.): a study of the problem and remedial action taken. *Water Research*, 17, 789–98.

Benton, C., Forbes, G.I., Paterson, G.M. *et al.* (1989) The incidence of waterborne and water associated disease in Scotland from 1945–1987. *Water Science and Technology*, 21(3), 125–9.

Berg, J.D. and Fiskdal, L. (1988) Rapid detection of total and faecal coliforms in water by enzymatic hydrolysis of 4-methylumbelliferone-ß-D-galactoside. *Applied and Environmental Microbiology*, 54, 2118–22.

Berger, P.S. (1992) Revised total coliform rule, in *Regulating Drinking Water*, (eds C.E. Gilbert and E.J. Calabrese), Lewis, Boca Raton, FL, pp. 161–6.

Bermundez, M. and Hazen, T.C. (1988) Phenotypic and genotypic comparison of *Escherichia coli* from pristine tropical waters. *Applied and Environmental Microbiology*, 54, 979–83.

Bhattacharya, S.K., Bhattacharya, M.K., Nair, G.B. *et al.* (1993) Clinical profile of acute diarrhoeal cases infected with the new epidemic strains of *V. cholerae* 0139: Designation of the disease as cholera. *Journal of Infection*, 27, 11–15.

Bisson, J.W. and Cabelli, V.J. (1979) Membrane filtration enumeration for *Clostridium perfringens*. *Applied and Environmental Microbiology*, 37, 55–66.

Bisson J.W. and Cabelli, V.J. (1980) *Clostridium perfringens* as a water pollution indicator. *Journal of the Water Pollution Control Federation*, 52, 241–8.

Bissonette, G.K., Jezeski, J.J., McFeters, G.A. and Stuart, D.G. (1975) Influence of environmental stress on enumeration of indicator bacteria from natural waters. *Applied and Environmental Microbiology*, 29, 186–94.

Bissonette, G.K., Jezeski, J.J., McFeters, G.A. and Stuart, D.G. (1977) Evaluation of recovery methods to detect coliforms in water. *Applied and Environmental Microbiology,* 33, 590–5.

Bitton, G. (1978) Survival of enteric viruses, in *Water Pollution Microbiology,* vol. 2, (ed. R. Mitchell), Wiley Interscience, New York, pp. 273–300.

Bitton, G. (1994) *Wastewater Microbiology.* Wiley, New York.

Blaser, M.J., Pollard, R.A. and Feldman, R.A. (1984) Shigella infections in the United States 1974–1980. *Journal of Infectous Diseases,* 147, 771–5.

Blaser, M.J., Taylor, D.N. and Feldman, R.A. (1982) Epidemiology of *Campylobacter jejuni* infections. *Epidemiology Reviews,* 5, 157–76.

Blaser, M.J., Smith, P.F., Wang Wen-Lan, L. and Hoff J.C. (1986) Inactivation of *Campylobacter jejuni* by chlorine and monochloramine. *Applied and Environmental Microbiology,* 51, 307–11.

Blewett, D.A., Wright, J.J., Casemore, D.P. *et al.* (1993) Infective dose size studies on *Cryptosporidium parvum* using gnotobiotic lambs. *Water Science and Technology,* 27, 61–3.

Block, J.C. (1992) Biofilms in drinking water distribution systems, in *Biofilms Science and Technology,* (eds L.F. Melo *et al.*), Kluwer, Amsterdam, pp. 469–85.

Bodhidatta, L., Echeverria, P., Hoge, C.W. *et al.* (1995) *Vibrio cholerae* 0139 in Thailand in 1994. *Epidemiology and Infection,* 114, 71–3.

Bolton, F.J., Hinchliffe, P.M., Coates, D. and Robertson, L. (1982) A most probable number method for estimating small numbers of *Campylobacter* in water. *Journal of Hygiene, Cambridge,* 89, 185–90.

Bonde, G.J. (1977) Bacterial indicators of water pollution, in *Advances in Advanced Aquatic Microbiology,* (eds M.R. Droop and H.W. Jamasdi), Academic Press Inc., London.

Borrego, J.J., Morinigo, M.M., de Vicente, A. *et al.* (1987) Coliphage as an indicator of faecal pollution in water. Its relationship with indicator and pathogenic organisms. *Water Research,* 21, 1473–80.

Borrego, J.J., Cornax, R., Morinigo, M. *et al.* (1990) Coliphage as an indicator of faecal pollution. Their survival and productive infectivity in natural aquatic environments. *Water Research,* 24, 111–16.

Borup, M.B. (1992) Presence–absence coliform monitoring has statistical limitations. *Journal of the American Water Works Association,* 84, 66–71.

Bradley, D.J. (1993) Human tropical diseases in a changing environment, in *Environmental Change and Human Health,* Wiley, Chichester, Ciba Foundation Symposium, 171, 146–70.

Brenner, N.P. and Rankin, C.C. (1990) New screening test to determine the acceptability of 0.45 μm membrane filters for analysis of water. *Applied and Environmental Microbiology,* 56, 54–64.

Brenner, N.P., Rankin, C.C., Roybal, Y.R. *et al.* (1993) New medium for the simultaneous detection of total coliforms and *E. coli* in water. *Applied and Environmental Microbiology,* 59, 3534–44.

Briscou, J. (1975) Yeasts and fungi in marine environments. *Société Française Mycologie Medicale Bulletin,* 4, 159–62.

Brodsky, M.H. and Schiemann, D.A. (1975) Influence of coliform source on evaluation of membrane filters. *Applied Microbiology,* 30, 727–30.

Brown, R.C. and Wade, T.L. (1984) Sedimentary coprostanol and hydrocarbon distribution adjacent to a sewage outfall. *Water Research,* 18, 621–32.

Browning, J.R. and Ives, D.G. (1987) Environmental health and the water distribution system: a case history of an outbreak of giardiasis. *Journal of the Institution of Water and Environmental Management*, 1, 55–60.

Bruce, D., Zochowski, W. and Fleming, G.A. (1980) Campylobacter infections in cats and dogs. *Veterinary Record*, 107, 200–1.

Buck, J.D. (1977) *Candida albicans*, in *Bacterial Indicators Health Hazards Associated With Water*, (eds A.W. Hoadley and B.J. Dutka), ASTM Special Technical Publication: 635, ASTM, Philadelphia, PA, pp. 139–47.

Buck, J.D. and Bubucis, P.M. (1978) Membrane filter procedure for enumeration of *Candida albicans* in natural water. *Applied Environmental Microbiology*, 35, 237–42.

Buckley, H.R. (1971) Fungi pathogenic for man and animals: 2. The subcutaneous and deep seated mycoses, in *Methods in Microbiology*, vol. 4., (ed. C. Booth), Academic Press, London, pp. 461–78.

Burke, V.J., Robinson, M., Gracey, M. *et al.* (1984a) Isolation of *Aeromonas hydrophila* from a metropolitan water supply: Seasonal correlation with clinical isolates. *Applied and Environmental Microbiology*, 48, 361–6.

Burke, V.J., Robinson, M., Gracey, M. *et al.* (1984b) Isolation of *Aeromonas* spp. from an unchlorinated domestic water supply. *Applied and Environmental Microbiology*, 48, 367–70.

Cabelli, V. (1978) New Standards for enteric bacteria, in *Water Pollution Microbiology*, vol. 2, (ed. R. Mitchell), Wiley-Interscience, New York, pp. 233–73.

Calabrese, J.P. and Bissonette, G.K. (1988) Modification of standard recovery media for enhanced detection of chlorine stressed coliform and heterotrophic bacteria. *Abstracts of the 89th Annual Meeting of the American Society for Microbiologists*, 89, 370.

Calabrese, J.P. and Bissonette, G.K. (1990) Improved membrane filtration method incorporating catalase and sodium pyruvate for detection of chlorine-stressed coliform bacteria. *Applied and Environmental Microbiology*, 56, 3558–64.

Caldwell, B.A. and Monta, R.A. (1988) *Sampling Regimes and Bacteriological Tests for Coliform Detection in Ground Water*, Project Summary EPA/600/82–87/083, US EPA, Cincinnati, OH.

Carrillo, M., Estrada, E. and Hazen, T.C. (1985) Survival and enumeration of the faecal indicators *Bifidobacteria adolescentis* and *E. coli* in a tropical rain forest watershed. *Applied and Environmental Microbiology*, 50, 468–76.

Carter, A.M., Pacha, R.E., Clark, G.W. and Williams, E.A. (1987) Seasonal occurrence of *Campylobacter* spp. in surface waters and their correlation with standard indicator bacteria. *Applied and Environmental Microbiology*, 53, 523–6.

Casemore, D. (1990) Epidemiological aspects of human cryptosporidiosis. *Epidemiology and Infection*, 104, 1–28.

Chamberlain, C.E. and Mitchell, R. (1978) A decay model for enteric bacteria in natural waters, in *Water Pollution Microbiology*, vol. 2, (ed. R. Mitchell), Wiley, New York, pp. 325–48.

Clark, D.L., Milner, B.B., Stewart, M.H. *et al.* (1991) Comparative study of commercial 4-methylumbelliferyl-ß-D-glucuronide preparations with the standard methods membrane filtration faecal coliform test for the detection of *E. coli* in water samples. *Applied and Environmental Microbiology*, 57, 1528–34.

Clark, J.A. (1968) A P–A test providing sensitive and inexpensive detection of coliforms, faecal coliforms and faecal streptococci in municipal drinking water supplies. *Canadian Journal of Microbiology*, 14, 13–18.

Clark, J.A. (1969) The detection of various bacteria indicative of water pollution by a P–A procedure. *Canadian Journal of Microbiology*, 15, 771–80.

Clark, J.A. (1980) The influence of increasing numbers of non indicator organisms by the membrane filtration and P–A tests. *Canadian Journal of Microbiology*, 26, 827–32.

Cochran, W.G. (1950) Estimation of bacterial densities by means of the most probable number. *Biometrics*, 5, 105–9.

Colquhoun, K.O., Timms, S. and Fricker, C.R. (1995) Detection of *E. coli* in potable water using direct impedance technology. *Journal of Applied Bacteriology*, 79, 635–9.

Contruvo, J.A. (1989) Drinking water standards and risk assessment. *Journal of the Institution of Water and Environmental Management*, 3, 6–12.

Cornax, R., Morinigo, M., Balebona, M.G. *et al.* (1991) Significance of several bacteriophage groups as indicators of sewage pollution in marine waters. *Water Research*, 25, 673–8.

Covert, T.C., Shadix, L.C., Rice, E.W. *et al.* (1989) Evaluation of autoanalysis colilert test for detection and enumeration of total coliforms. *Applied and Environmental Microbiology*, 55, 2443–7.

Covert, T.C., Rice, E.W., Johnson, B.A. *et al.* (1992) Comparative defined-substrate coliform tests for the detection of *E. coli* in water. *Journal of the American Water Works Association*, 84, 98–104.

Cowburn, J.K., Goodall, T., Fricker, E.J. *et al.* (1994) A preliminary study of the use of Colilert for water quality monitoring. *Letters in Applied Microbiology*, 19, 50–2.

Craun, G.F. (1977) Waterborne outbreaks. *Water Pollution Control Federation*, 49, 1268.

Craun, G.F. (1986) *Waterborne Diseases in the United States*, CRC Press, Boca Raton, FL.

Craun, G.F. (1988) Surface water supplies and health. *Journal of the American Water Works Association*, 80, 40–50.

Craun G.F. (1991) Cause of waterborne outbreaks in the United States. *Water Science and Technology*, 24, 17–20.

Craun, G.F. (1992) Waterborne disease in the USA: causes and prevention. *World Health Statistics Quarterly*, 45, 192–9.

Craun, G.F., Batik, O. and Pipes, W.O. (1983) Routine coliform monitoring and waterborne disease outbreaks. *Journal of Environmental Health*, 45, 227–30.

Crewe, S.M. (1967) Worm eggs found in gull droppings. *Annals of Tropical Medicine and Parasitology*, 61, 358.

Crowley, F.W. and Packham, R.F. (1993) Water Treatment in Europe and North America. *Journal of the Institution of Water and Environmental Management*, 7, 81–9.

Cruz, J.R., Caceres, P., Cano, F. *et al.* (1990) Adenovirus types 40 and 41 and rotaviruses associated with diarrhea in children from Guatemala. *Journal of Clinical Microbiology*, 28, 1780–4.

Cubitt, D.W. (1991) A review of the epidemiology and diagnosis of waterborne viral infections: II. *Water Science and Technology*, 24, 197–203.

Dadswell, J.V. (1990a) Microbiological aspects of water quality. *Journal of the Institution of Water and Environmental Management*, 4, 515–23.

Dadswell, J.V. (1990b) Will privatisation affect water microbiology? *PHLS Microbiology Digest*, 7, 96–100.

Dart, R.K. and Stretton, R.J. (1977) *Microbial Aspects of Pollution Control*, Elsevier, Oxford.

Dennis, J.M. (1959) 1955–1956, infectious hepatitis epidemic in Delhi, India. *Journal of the American Water Works Association*, 31, 1288–98.

Department of the Environment (1994a) The microbiology of water 1994: Part 1. Drinking water. *Reports on Public Health and Medical Subjects No. 71. Methods for the Examination of Water and Associated Materials*, HMSO, London.

Department of the Environment (1994b) *Drinking Water 1993. A Report by the Chief Inspector, Drinking Water Inspectorate*, HMSO, London.

Department of the Environment (1995) *Drinking Water 1994. A Report by the Chief Inspector, Drinking Water Inspectorate*, HMSO, London.

Department of the Environment and Department of Health (1990) *Cryptosporidium in Water Supplies. Report of a Group of Experts*, HMSO, London.

Department of the Environment, Department of Health and Social Security, and Public Health Laboratory Service (1983). The bacteriological examination of drinking water supplies 1982. *Reports on Public Health and Medical Subjects No. 71. Methods for the Examination of Water and Associated Materials*, HMSO, London.

Dev, V.J., Main, M. and Gould, I. (1991) Waterborne outbreak of *E. coli* 0157. *Lancet*, 337, 1412.

Devriese, L.A., Laurier, L., de Herdt, P. and Hasebrouck, F. (1992) Enterococcal and streptococcal spp. isolated from faeces of calves, young cattle and dairy cows. *Journal of Applied Bacteriology*, 72, 29–31.

De Zuane, J. (1990) *Handbook of Drinking Water Quality Standards and Controls*, Van Nostrand Reinhold, New York.

Dhillon, T.S., Dhillon, E.K., Chau, H.C. *et al.* (1976) Studies on bacteriophage distribution: virulent and temperate bacteriophage content of mammalian faeces. *Applied and Environmental Microbiology*, 32, 68–74.

Dionisio, L.P.C. and Borrego, J.J. (1995) Evaluation of media for the enumeration of faecal streptococci from natural water samples. *Journal of Microbiological Methods*, 23, 183–203.

Divizia, M., Morace, G., Gabrieli, R. *et al.* (1993) Application of the PCR technique to the detection of hepatitis A in the environment. *Water Science and Technology*, 27, 223–7.

Dufour, A.P., Strickland, E.R. and Cabelli, V.J. (1981) Membrane filter method for enumerating *E.coli. Applied and Environmental Microbiology*, 41, 1152–8.

Du Moulin, G.C. and Stottmeier, K.D. (1986) Waterborne mycobacteria: an increasing threat to health. *American Society of Microbiology News*, 1986, 525–9.

Dutka, B.J. (1973) Coliforms are an inadequate index of water quality. *Journal of Environmental Health*, 36, 39–46.

Dutka, B.J. (1978) *Methods for Microbial Analysis of Waters, Wastewaters and Sediments*, Inland Waters Directorate, Environment Canada, Ontario, pp. II.44–7.

Dutka, B.J. (1979) Microbiological indicators, problems and potential of new microbial indicators of water quality, in *Biological Indicators of Water Quality*, (eds A. James and L. Evison), Wiley, Chicester, pp. 18/1–21.

Dutka, B.J. and Bell, J.B. (1973) Isolation of *Salmonellae* from moderately polluted waters. *Journal of the Water Pollution Control Federation*, 45, 316–24.

Dutka, B.J. and El-Shaarawi, A. (1975) Relationship between various bacterial populations and coprostanol and cholesterol. *Canadian Journal of Microbiology*, 21, 1386–98.

Dutka, B.J. and Kwan, K. (1980) Bacterial die-off and stream transport studies. *Water Research*, 14, 909–15.

Dutka, B.J. and Tobin, S.E. (1976) Study of the efficiency of four procedures for enumerating coliforms in water. *Canadian Journal of Microbiology*, 22, 630–5.

Dutka, B.J., Chan, A.S. and Coburn, J. (1974) Relationship between bacterial indicators of water pollution and faecal sterols. *Water Research*, 8, 1047–55.

Dutka, B.J., Jackson, M.J. and Bell, J.B. (1974) Comparison of autoclave and ethylene oxide sterilised membrane filters used in water quality studies. *Applied Microbiology*, 28, 474–480.

EC (1980) Council Directive relating to the quality of water intended for human consumption (80/778/EEC). *Official Journal of the European Community*, L229, (30.08.80), 11–29.

Edberg, S.C., Allen, M.J. and Smith, B.D. (1989) National field evaluation of a defined substrate method for the simultaneous enumeration of total coliforms and *E. coli* from drinking water: comparison with standard methods P–A technique. *Applied and Environmental Microbiology*, 55, 1003–8.

Edberg, S.C., Allen, M.J., Smith D.B. and the National Collaborative Study (1988) National field evaluation of a defined substrate method for the simultaneous enumeration of total coliforms and *E. coli* from drinking water: Comparison with the standard method multiple tube method. *Applied and Environmental Microbiology*, 54, 1595–601.

Edberg, S.C., Allen, M.J., Smith, D.B. and King, N.J. (1990) Enumeration of total coliforms and *E. coli* from source waters by the defined substrate technique. *Applied and Environmental Microbiology*, 56, 366–9.

Ellis, K.V. (1989) *Surface Water Pollution and its Control*. Macmillan, London.

ElShaarawi, A.H. and Pipes, W.O. (1983) Enumeration and statistical interference, in *Bacterial Indicators of Pollution*, (ed. W.O. Pipes), CRC Press, Boca Raton, FL, pp. 43–66.

Enriquez, C.E. and Gerba, C.P. (1995) Concentration of enteric Adenovirus 40 from tap, sea and waste water. *Water Research*, 29, 2554–60.

Esrey, S.A. and Habicut, J.P. (1986) Epidemiolgic evidence for health benefits from improved water and sanitation in developing countries. *Epidemiological Reviews*, 8, 117–28.

Esterman, A., Roder, D.M., Cameron, A.S. *et al.* (1984) Determinants of the microbial chracteristics of South Australian swimming pools. *Applied and Environmental Microbiology*, 47, 325–8.

EU (1995) Proposal for a Council Directive concerning the quality of water intended for human consumption. Com (94) 612 Final. *Official Journal of the European Union*, 131, (30.05.95), 5–24.

Evans, T.M., Seidler, R.J., and Le Chevallier, M.W. (1981) Impact of verification media and resuscitation on the accuracy of the membrane filtration total coliform enumeration technique. *Applied and Environmental Microbiology*, 41, 1144–57.

Evans, T.M., Le Chevallier, M.W., Waarvick, C.E. and Seidler, R.J. (1981a) Coliform species recovered from untreated surface water and drinking water by the membrane filter, standard and modified most probable number technique. *Applied and Environmental Microbiology*, 41, 657–63.

Evans, T.M., Waarvick, C.E., Seidler, R.J. and Le Chevallier, M.W. (1981b) Failure of the most probable number technique to detect coliforms in drinking water and raw water supplies. *Applied and Environmental Microbiology*, 41, 130–58.

Evison, L.M. and James, A. (1973) A comparison of the distribution of intestinal bacteria in Britain and East African water sources. *Journal of Applied Bacteriology*, 36, 109–18.

Evison, L.M. and Morgan, S. (1978) Further studies on *Bifidobacteria* as indicators of faecal pollution in water. *Progress in Water Technology*, 10, 341–50.

Faubert, G.M. (1988) Evidence that giardiasis is a zoonosis. *Parasitology Today*, 4, 66–8.

Fawell, J.K. and Miller, D.G. (1992) U.K. drinking water. A European comparison. *Journal of the Institution of Water and Environmental Management*, 6, 726–33.

Fayer, R. and Ungar, B.L. (1986) *Cryptosporidium* spp. and cryptosporidiosis. *Microbiological Reviews*, 50, 458–83.

Feacham, R.G., Bradley, D.J. Gavelick, H. and Mara, D.D. (1983) *Sanitation and Disease: Health Aspects of Excreta and Wastewater Management*, Wiley, Chichester.

Federal Register (1988) *National Primary Drinking Water Regulation: Proposed Water Rule*: Federal Register 16348–58 (Proposed Rule).

Federal Register (1989) *Drinking Water, National Primary Drinking Water Regulations. Total Coliforms Final Rule (including E. coli and faecal coliforms)*. Federal Register (54) 27544–7.

Feng, P.C.S. and Hartman, P.A. (1982) Fluorogenic assay for immediate confirmation of *E. coli*. *Applied and Environmental Microbiology*, 43, 1320–9.

Fennell, H., James, D.B., and Morris, J. (1974) Pollution of a storage reservoir by roosting gulls. *Water Treatment and Examination* 23, 5–24.

Finelli, L., Crayne, E., Dalley, E. *et al*. (1994) Enhanced detection of sporadic *E. coli* 0157:H7 infections New Jersey, 1994. *MMWR*, 44, 417–18.

Firstenberg-Eden, R. and Eden, G. (1984) *Impedance Microbiology*, Research Studies Press, Letchworth.

Flanagan, P.J. (1990) *Parameters of water quality: Interpretation and standards*, Environmental Research Unit, Dublin.

Fox, J.G., Ackerman, J.I. and Newcomer, C.E. (1983) Ferret as a potential reservoir for human campylobacteriosis. *American Journal of Veterinary Research*, 44, 1049–52.

Fox, L.J., Keller, N. and Schothoist, M. van (1988) The use and misuse of quantitative determinations of Enterobacteriaceae in food microbiology. *Journal of Applied Bacteriology* (Symposium Supplement), 2375–95.

Fox, J.G., Zanotti, S. and Jordan, H.V. (1981) The hamster as a reservoir of *Campylobacter fetus* subspecies *jejuni*. *Journal of Infectious Diseases*, 143, 856.

Frampton, E.W. and Restaino, L. (1993) A review: methods for *E. coli* identification in food. *Journal of Applied Bacteriology*, 74, 223–33.

Freeman, R., Sisson, P.E., Jenkins, D.R. *et al.* (1995) Sporadic isolates of *Escherichia coli* 0157:H7 investigated by pyrolysis mass spectrometry. *Epidemiology and Infection*, 114, 433–40.

Freier, T.A. and Hartman, P.A. (1987) Improved membrane filtration media for enumeration of total coliforms and *E. coli* from sewage and surface waters. *Applied and Environmental Microbiology*, 53, 1246–50.

Fricker, E.J. and Fricker, C.P. (1994) Application of the polymerase chain reaction to the identification of *E. coli* and coliforms in water. *Letters in Applied Microbiology*, 19, 44–7.

Fujioka, R.S. and Shizumura, L.K. (1985) *Clostridium perfringens*, a reliable indicator of stream water quality. *Journal of the Water Pollution Control Federation*, 57, 986–92.

Furness, M.L. and Holmes, P. (1987) Bacteriological monitoring of water with 'Jungle Kits'. *Journal of the Institution of Water and Environmental Management*, 1, 227–50.

Galbraith, N.S., Barnett, N.J. and Stanwell-Smith, R. (1987) Water and disease after Croydon: A review of waterborne and water associated disease in the United Kingdom, (1937–1986). *Journal of the Institution of Water and Environmental Management*, 1, 7–21.

Gale, P. and Broberg, P.J. (1993) Evaluation of a rapid, defined substrate technique method for enumerating total coliforms and *E. coli* in chlorinated drinking water. *Letters in Applied Microbiology*, 17, 200–3.

Gauci, V. (1991) Enumeration of faecal streptococci in seawater, in *Development and Testing of Sampling and Analytical Techniques for Monitoring of Marine Pollutants (Activity A):* Final reports on selected microbiological projects, MAP Technical Reports series No. 54, UNEP, Athens, pp. 47–59.

Geldenhuys, J.C. and Pretorius, P.D. (1989) The occurrence of enteric viruses in polluted water, correlation to indicator organisms and factors influencing their numbers. *Water Science and Technology*, 21, 105–9.

Geldreich E.E. (1970) Applying bacteriological parameters to recreational water quality. *Journal of the American Water Works Association*, 62, 113–20.

Geldreich, E.E. (1991) Microbial water quality concerns for water supply use. *Environmental Toxicology and Water Quality*, 6, 209–23.

Geldreich, E.E. (1992) Visions of the future in drinking water microbiology. *Journal of the New England Water Works Association*, 56, 1–8.

Geldreich, E.E. (1996) *Microbial Quality of Water Supply in Distribution Systems.* Lewis Publishers, Boca Raton, FL.

Geldreich, E.E. and Kennedy, H. (1982) The cost of microbiological monitoring, in *Bacterial Indicators of Pollution*, (ed. W.O. Pipes), CRC Press, Boca Raton, FL, pp. 131–48.

Geldreich, E.E. and Kenner, B.A. (1969) Concepts of faecal streptococci in stream pollution. *Journal of the Water Pollution Control Federation*, 41, 336–51.

Geldreich, E.E. and Rice, E.W. (1987) Occurrence, significance and detection of *Klebsiella* in water systems. *Journal of the American Water Works Association*, 79, 74–80.

Geldreich, E.E., Nash, H.D., Reasoner, D.J. and Taylor, R.H. (1972) The necessity of controlling bacterial populations in potable waters: community water supply. *Journal of the American Water Works Association*, 64, 596–602.

Geldreich, E.E., Fox, K.R., Goodrich, J.A. *et al.* (1992) Searching for a water supply connection in the Cabool, Missouri disease outbreak of *Escherichia coli* 0157:H7. *Water Research*, 26, 1127–37.

Gerba, C.P. and Rose, J.B. (1990) Viruses in source and drinking water, in *Drinking Water Microbiology*, (ed. G.A. McFeters), Springer-Verlag, New York, pp. 380–96.

Gerba, C.P., Margolin, A.B. and Hewlett, J.M. (1989) Application of gene probes to virus detection in water. *Water Science and Technology*, 21, 147–54.

Gerba, C.P., Goyal, S.M., LaBelle, R.L. *et al.* (1979) Failure of indicator bacteria to reflect the occurrence of Enteroviruses in marine waters. *American Journal of Public Health*, 69, 1116–9.

Girones, R., Allard, A., Wadell, G. and Jofre, J. (1993) Application of PCR to the detection of Adenoviruses in polluted waters. *Water Science and Technology*, 27, 235–43.

Gondrosen, B., Melby, K., Gregusson, S. and Dahl, O.P. (1985) A waterborne outbreak of campylobacter enteritis in the subarctic region of Norway, in *Campylobacter*, vol. 3, (eds A.D. Pearson, M.B. Skirrow, H. Lior and R. Rowe), Public Health Laboratory Service, London, p. 277.

Gould, D.J. (1977) *Gull Droppings and their Effect on Water Quality*, Technical Report TR37. Water Research Centre, Stevenage.

Grabow, W.O. (1986) Indicator systems for assessment of the viriological safety of drinking water. *Water Science and Technology*, 18, 159–68.

Grabow, W.O. and Du Preez, M. (1979) Comparison of M–ENDO LES, McConkey and Teepol Media for membrane filtration counting of total bacteria in water. *Applied and Environmental Microbiology*, 38, 351–8.

Grabow, W.O., Burger, J.S. and Nupen, E.M. (1980) Evaluation of acid fast bacteria, *Candida albicans*, enteric viruses and conventional indicators for monitoring wastewater reclamation systems. *Progress in Water Technology*, 12, 803–17.

Grant, M.A. (1996) Hatch Co., Iowa, personal communication.

Gray, N.F. (1992) *Biology of Wastewater Treatment*, Oxford University Press, Oxford.

Gray, N.F. (1994) *Drinking Water Quality: Problems and Solutions*, Wiley, Chichester.

Gray, S.F., Gunnell, D.J. and Peters, T.J. (1994) Risk factors for giardiasis: a case-control study in Avon and Somerset. *Epidemiology and Infection*, 113(1), 95–102.

Green, B.L., Clausen, E.M and Litsky, W. (1975) Comparison of the new Millipore HC with conventional membrane filters for the enumeration of faecal coliform bacteria. *Applied Microbiology*, 30, 697–9.

Green, B.L., Clausen, E.M. and Litsky, W. (1977) Two temperature membrane filter for enumerating faecal coliform bacteria from chlorinated effluents. *Applied and Environmental Microbiology*, 33, 1259–64.

Greene, V.W. (1982) Public health, quality of life and lifestyle hazards, in *Bacterial Indicators of Pollution*, (ed. W.O. Pipes), CRC Press, Boca Raton, FL, pp. 149–66.

Grimalt, J.O., Fernandez, P., Boyona, J.P. and Albraiges, J. (1990) Assessment of fecal sterols and ketones as indicators of urban sewage inputs to coastal waters. *Environmental Science and Technology*, 24, 357–62.

Hancock, D.D., Besser, T.E., Kinsel, M.L. *et al.* (1994) The prevalence of *Escherichia coli* 0157:H7 in dairy and beef cattle in Washington State. *Epidemiology and Infection*, 113, 199–208.

Hasset, J.P. and Lee, G.F. (1977) Sterols in natural waters and sediments. *Water Research*, 11, 983–9.

Havelaar, A.H. and Hogeboom, W.M. (1984) A method for the enumeration of male-specific bacteriophages in sewage. *Journal of Applied Bacteriology*, 56, 439–47.

Havelaar, A.H. and Hogeboom, W.M. (1988) F-specific RNA-bacteriophages as model viruses in water hygiene: Ecological aspects. *Water Science and Technology*, 20, 399–407.

Havelaar, A.H., Furose, K. and Hogeboom, W.M. (1986) Bacteriophages and indicator bacteria in human and animal faeces. *Journal of Applied Bacteriology*, 60, 255–62.

Havelaar, A.H., Hogeboom, W.M., Furose, K. *et al.* (1990) F-specific RNA bacteriophage and sensitive host strains in faeces and wastewater of human and animal origin. *Journal of Applied Bacteriology*, 69, 30–7.

Hayes, C.R. (1989) Microbiological quality control in the provision of safe drinking water. *Water Science and Technology*, 21, 559–66.

Hayes, M. and Cooper, R.A. (1994) Cryptosporidiosis hidden in name and nature. *Safety and Health Practitioner*, August, 16–20.

Hejkal, T.W., Keswick, B., La Belle, R.L. *et al.* (1982) Viruses in a community water supply associated with an outbreak of gastroenteritis and infectious hepatitis. *Journal of the American Water Works Association*, 74, 318–21.

Herwaldt, B.L., Craun, G.F. Stokes, S.L. and Juranek, D.D. (1992) Outbreaks of water disease in the United States: 1989–1990. *Journal of the American Water Works Association*, 84, 129–35.

Hickling, R.A.O. (1977) Inland wintering of gulls in England and Wales, 1973. *Bird Study*, 24, 78–88.

Hill, G.A. and Grimes, D.J. (1984) Seasonal study of a freshwater lake and migratory waterfowl for *Campylobacter jejuni*. *Canadian Journal of Microbiology*, 30, 845–9.

HMSO (1989) *The Water Supply (Water Quality) Regulations*, 1989, Statutory Instrument 1989/1147. HMSO, London.

Hofstra, H. and Huis In't Veld, J.H.J. (1989) Methods for the detection and isolation of *E. coli* including pathogenic strains. *Journal of Applied Bacteriology* (Symposium Supplement), 1975–2125.

Holmes, B. and Costas, M. (1992) Identification and typing of Enterobacteriaceae by computerised methods, in *Identification Methods in Applied and Environmental Microbiology*, (eds R.G. Board, D. Jones and F.A. Skinner), Blackwell Scientific Publications, Oxford, pp. 127–51.

Holt, J.G., Krieg, N.R., Sneath, P.H.A. *et al.* (eds) (1993) *Bergey's Manual of Determinative Bacteriology*, 9th edn, Williams and Wilkins, Baltimore, MD.

Holt, P.E. (1980) Incidence of campylobacter, salmonella and shigella infections in dogs in an industrial town. *Veterinary Record*, 107, 254.

Hood M.A., Mois, G.E. and Blake, N.J. (1983) Relationship between faecal coliforms, *E. coli* and *Salmonella* spp. in shellfish. *Applied and Environmental Microbiology*, 45, 122–6.

Hornick, R.B. (1985) Selective primary health care: strategies for control of disease in the developing world. XX: Typhoid fever. *Review of Infectious Diseases*, 7, 536–46.

Huck, P.M. (1990) Measurement of biodegradable organic matter and bacterial growth potential in drinking water. *Journal of the American Water Works Association*, 82, 78–86.

Huck, P.M., Fedovak, P.M. and Anderson, W.B. (1991) Formation and removal of assimilable organic carbon during biological treatment. *Journal of the American Water Works Association*, 83, 69–74.

Hussong, D.J, Damare, J.M., Limpert, R.J. *et al.* (1979) Microbial impact of Canada Geese (*Branta canadensis*) and Whistling Swans (*Cygnus columbianus*) on aquatic ecosystems. *Applied and Environmental Microbiology*, 37, 14–20.

Hussong, D.J., Darnone, M., Weiner, R.M. and Colwell, R.R. (1981) Bacteria associated with faecal pollution MPN coliform tests, results for shellfish and estuaries. *Applied and Environmental Microbiology*, 41, 35–45.

Hutchison, D., Weaver, R.H. and Scherago, M. (1943) The incidence and significance of microorganisms antagonistic to *E. coli. Journal of Bacteriology*, 45, 29.

Irving, T.E., Stanfield, G. and Hepburn, B.W.T. (1989) Electrical methods for water quality testing, in *Rapid Microbiological Methods for Foods, Beverages and Pharmaceuticals* (eds C.J. Stannard, S.B. Petit, and F.A. Skinner), Society for Applied Bacteriology, Technical Series No. 25, Blackwell Science, Oxford, pp. 119–30.

Jacobs, N.J., Zeigler, W.L., Reed, F.C. *et al.* (1986) Comparison of membrane filter, multiple fermentation tube and P–A techniques for detecting coliforms in small community water systems. *Applied and Environmental Microbiology*, 51, 1007–12.

Jay, J.M. (1991) *Modern Food Microbiology*, Chapman & Hall, New York.

Jenkins, P.A. (1991) Mycobacterium in the environment. *Journal of Applied Bacteriology* (Symposium Supplement), 137–41.

Jephcote, A.E., Begg, N. and Baker, I. (1986) An outbreak of Giardiasis associated with mains water in the United Kingdom. *Lancet*, (8483), 730–2.

Jofre, J., Bosch, A., Lucena, F. *et al.* (1987) Evaluation of *Bacteroides fragilis* bacteriophage as indicators of the viriological quality of water. *Water Science and Technology*, 18, 167–73.

Johnson, D.W., Pienazek, N.J. and Rose, J.B. (1993) DNA probe hybridisation and PCR detection of *Cryptosporidium* compared to immunofluorescence assay. *Water Science and Technology*, 27, 77–84.

Johnson, D.W., Pienazek, N.J., Griffin, D.W. *et al.* (1995) Development of a PCR protocol for sensitive detection of *Cryptosporidium* oocysts in water samples. *Applied and Environmental Microbiology*, 61, 3849–55.

Johnstone, D.W. and Horan, N.J (1994) Standards, costs and benefits: an international perspective. *Journal of the Institution of Water and Environmental Management*, 8, 450–8.

Jones, K. (1994) Inside science: 73. Waterborne diseases. *New Scientist*, **143**(1933) 1–4.

Jones, K. and Telford, D. (1991) On the trail of the seasonal microbe. *New Scientist*, 130, 36.

Jones, R.E. (1978) *Heavy Metals in the Estuarine Environment*, Technical Report 78, Water Research Centre, Stevenage.

Joret, J.C., Cervantes, P., Levi, Y. *et al.* (1989) Rapid detection of *E. coli* in water using monoclonal antibodies. *Water Science and Technology*, 21, 161–7.

Joseph, C.A., Watson, J.M., Harrison, T.G. and Bartlett, C.L.R. (1994) Nosocomial Legionnaires' disease in England and Wales 1980–92. *Epidemiology and Infection*, 112, 329–46.

Kansouzidou, A., Paneri, V., Bourelakis, D. and Danielides, B.D. (1991) Enterohaemorrhagic *Escherichia coli* O157:H7. Do they exist in Greece? *Applied Clinical Microbiology*, 6, 30–4.

Kaper, J.H., Sayler, G.S., Baldini, M. and Colwell, R. (1979) Ecology, serology and enterotoxin production of *Vibrio cholerae* in Chesapeake Bay. *Applied and Environmental Microbiology*, 33, 829–35.

Keswick, B.H., Shatterwhite, T.K., Johnson, P.C. *et al.* (1985) Inactivation of the Norwalk virus in drinking water by chlorine. *Applied and Environmental Microbiology*, 50, 261–4.

Kfir, R., Du Preez, M. and Genthe B. (1993) The use of monoclonal antibodies for the detection of faecal bacteria in water. *Water Science and Technology*, 27, 257–60.

Knittel, M.D., Seidler, R.J., Eby, C. and Cabe, L.M. (1977) Colonisation of the botanical environment by *Klebsiella* isolates of pathogenic origin. *Applied and Environmental Microbiology*, 34, 557–63.

Kooij, D. van der (1990) Assimilable organic carbon (AOC) in drinking water, in *Drinking Water Microbiology* (ed. G.A. McFeters), Springer–Verlag, New York, pp. 57–87.

Korich, D.G., Mead, J.R., Madore, M.S. *et al.* (1990) Effects of chlorine, chlorine dioxide, chlorine and monochloramine on *Cryptosporidium parvum* oocyst viability. *Applied and Environmental Microbiology*, 56, 1423–8.

Kott, Y., Rose, N., Sperber, S. and Betzer, N. (1974) Bacteriophage as viral pollution indicators. *Water Research*, 8, 167–71.

Kurtz, J.B. and Lee, T.W. (1978) Astrovirus gastroenteritis: age distribution of antibody. *Medical Microbiology and Immunology*, 166, 277–30.

La Belle, R.L., Gerba, C.P., Goyal, S.M. *et al.* (1980) Relationships between environmental factors, bacterial indicators and the occurrence of enteric viruses in estuarine sediments. *Applied and Environmental Microbiology*, 39, 588–96.

Lamka, K.G., Le Chevallier, M.W. and Seidler, R.J. (1980) Bacterial contamination of drinking water supplies in a modern rural neighbourhood. *Applied and Environmental Microbiology*, 39, 734–8.

Lang, A.L., Tsai, Y.L., Mayer, C.L. *et al.* (1994) Multiplex PCR for detection of the heat labile toxin gene and Shiga-like toxin I and II genes in *Escherichia coli* isolated from natural waters. *Applied and Environmental Microbiology*, 60, 3145–9.

Lantz, N. and Hartman, P.A. (1976) Time released capsule method for coliform enumeration. *Applied and Environmental Microbiology*, 32, 716–22.

Lavoie, M.C. (1983) Identification of strains isolated as total and faecal coliforms and a comparison of both groups as indicators of faecal pollution in tropical climates. *Canadian Journal of Microbiology*, 29, 689–93.

Le Chevallier, M.W. (1990) Coliform regrowth in drinking water: a review. *Journal of the American Water Works Association*, 82, 74–86.

Le Chevallier, M.W., Babcock, T.M. and Lee, R.G. (1987) Examination and characterisation of distribution system biofilms. *Applied and Environmental Microbiology*, 53, 2714–24.

Le Chevallier, M.W., Cameron, S.C. and McFeters, G.A. (1983) New medium for improved recovery of coliform bacteria from drinking water. *Applied and Environmental Microbiology*, 45, 484–92.

Le Chevallier, M.W., Cawthon, C.P. and Lee, R.G. (1988a) Factors promoting survival of bacteria in chlorinated water supplies. *Applied and Environmental Microbiology*, 54, 649–54.

Le Chevallier, M.W., Cawthon, C.P. and Lee, R.G. (1988b) Inactivation of biofilm bacteria. *Applied and Environmental Microbiology*, 54, 2492–9.

Le Chevallier, M.W., Evans, T.M. and Seidler, R.J. (1981) Effect of turbidity on chlorination and bacterial persistence in drinking water. *Applied and Environmental Microbiology*. 42, 159–67.

Le Chevallier, M.W., Jakonski, P.E. and McFeters, G.A. (1984) Evaluation of M-T7 Agar as a faecal coliform medium. *Applied and Environmental Microbiology*, 48, 371–5.

Le Chevallier, M.W., Norton, W.D. and Lee, R.G. (1991a) Occurrence of *Giardia* and *Cryptosporidium* spp. in surface water supplies. *Applied and Environmental Microbiology*, 57, 2610–6.

Le Chevallier, M.W., Norton, W.D. and Lee, R.G. (1991b) *Giardia* and *Cryptosporidium* spp. in filtered drinking water supplies. *Applied and Environmental Microbiology*, 57, 2617–21.

Le Chevallier, M.W., Schulz, W. and Lee, R.G. (1991) Bacterial nutrients in drinking water. *Applied and Environmental Microbiology*, 57, 857–62.

Le Chevallier M.W., Seidler, R.J. and Evans, T.M. (1980) Enumeration and characterisation of SPC bacteria in chlorinated and raw water supplies. *Applied and Environmental Microbiology*, 40, 922–30.

Leclerc, H., Mossel, D.A.A., Trinel, P.A. and Gavini, F. (1976) A new test for faecal contamination, in *Bacterial Indicators – Health Hazards Associated with Water*, (eds A.W. Hoadley and B.J. Dutka), ASTM Publication: 635, ASTM, Philadelphia, PA.

Lee, J.L., Lightfoot, N.F. and Tillett, H.E. (1995) An evaluation of presence–absence techniques for coliform organisms and Escherichia coli 1995. *Methods for the Examination of Waters and Associated Materials*, Department of the Environment, HMSO, London.

Lee, J.V., Bashford, D.J., Donovan, T.J. *et al.* (1982) The incidence of *Vibrio cholerae* in water, animals and birds in Kent, England. *Journal of Applied Bacteriology*, 52, 281–8.

Lee, T.W. and Kurtz, J.B. (1994) Prevalence of human astrovirus serotypes in the Oxford Region 1976–92, the evidence for two new serotypes. *Epidemiology and Infection*, 112, 187–93.

Levin, M.A. and Resnick, I.G. (1981) Quantitative procedure for enumeration of *Bifidobacteria. Applied and Environmental Microbiology*, 42, 427–32.

Lewis, C.M. and Mak, J.L. (1989) Comparison of membrane filtration and auto-analysis colilert P–A techniques for analysis of total coliforms and *E. coli* in drinking water samples. *Applied and Environmental Microbiology*, 55, 3091–94.

Lin, S.D. (1976) A membrane filter method for recovery of faecal coliforms in chlorinated sewage effluents. *Applied and Environmental Microbiology*, 32, 300–2.

Lin, S.D. (1977) Comparison of membranes for faecal coliform recovery in chlorinated effluents. *Journal of the Water Pollution Control Federation*, 49, 2255–64.

Lin, S.D. (1985) *Giardia lamblia* and water supply. *Journal of the American Water Works Association*, 77, 40–7.

Liu, Z., Stout, J.E., Tedesco, L. *et al.* (1995) Efficacy of ultraviolet light in preventing *Legionella* colonization of a hospital water distribution system. *Water Research*, 29, 2275–80.

Loche, A. and Mach, L. (1988) Identification of HIV infected seronegative individuals by a direct diagnostic·test based on hybridization to amplified viral DNA. *Lancet*, 2, 418–21.

Lucena, F., Finance, C., Jofre, J. *et al.* (1982) Viral pollution determination of superficial waters (river water and sea water) from the urban area of Barcelona (Spain). *Water Research*, 16, 173–7.

Lynch, J.M. and Poole, N.J. (eds) (1979) *Microbial Ecology: a Conceptual Approach*, Blackwell, Oxford.

Madore, M.S., Rose, J.B., Gerba, C.P. *et al.* (1987) Occurrence of *Cryptosporidium* oocysts in sewage effluents and selected surface waters. *Journal of Parasitology*, 73, 702–5.

Maguire, H.C., Holmes, E., Hollyer, J. *et al.* (1995) An outbreak of cryptosporidiosis in south London. What value the *p* value? *Epidemiology and Infection*, 115, 279–87.

Mahbubani, M.H., Bej, A.K., Perlin, M. *et al.* (1991) Detection of *Giardia* cysts by using the polymerase chain reaction and distinguishing live from dead cysts. *Applied and Environmental Microbiology*, 57, 3456–61.

Manafi, M., Kneifel, W. and Bascomb, S. (1991) Fluorogenic and chromogenic substrates used in bacterial diagnostics. *Microbiological Reviews*, 55, 335–8.

Mancini, J.L. (1978) Numerical estimates of coliform mortality rates under various conditions. *Journal of the Water Pollution Control Federation*, 50, 2477–84.

Manja, K.S., Maurya, M.S. and Rao, K.M. (1992) A simple field test for the detection of faecal pollution in drinking water. *Bulletin of the World Health Organisation*, 60, 797–801.

Mara, D.D. (1974) *Bacteriology for Sanitary Engineers*, Churchill Livingstone, Edinburgh.

Mara, D.D. and Oragui, J.I. (1981) Occurrence of *Rhodococcus coprophilus* and associated actinomycetes in faeces, sewage and freshwater. *Applied and Environmental Microbiology*, 42, 1037–42.

Mara, D.D. and Oragui J.I. (1983) Sorbitol-fermenting *Bifidobacteria* as species indicators of human faecal pollution. *Journal of Applied Bacteriology*, 55, 349–57.

Marcola, B., Watkins, J. and Riley, A. (1981) The isolation and identification of thermotolerant *Campylobacter* spp. from sewage and river waters. *Journal of Applied Bacteriology*, 51, xii–xiv.

Marrie, T., Green, P., Burbridge, S. *et al.* (1994) *Legionellaceae* in the potable water of Nova Scotia hospitals and Halifax residences. *Epidemiology and Infection*, 112, 143–50.

Masters, G.A. (1991) *Introduction to Environmental Engineering and Science*, Prentice Hall, Englewood Cliffs, NJ.

Mates, A. and Shaffer, M. (1989) Membrane filter differentiation of *E. coli* from coliforms in the examination of water. *Journal of Applied Bacteriology*, 67, 343–6.

Mathieu, L., Block, J.C., Dutang, M. *et al.* (1994) Control of biofilm accumulation in drinking water distribution systems. *Water Supply*, 11, 365–76.

Maule, A. (1996) CAMAR, Porton Down, personal communication.

Maule, A., Vagost, D. and Block, J.C. (1991) Microbiology of distribution networks for drinking water supplies, in *Microbiological Analysis in Water Distribution Networks*, Ellis Horwood, Chichester, pp. 11–31.

McCarthy, S.C., Standridge, J.H. and Stasiak, M. (1992) Evaluating a commercially available defined substrate test for recovery of *E. coli*. *Journal of the American Water Works Association*, 84, 91–7.

McCrady, H. (1915) The numerical interpretation of fermentation tube results. *Journal of Infectious Diseases*, 17, 183–212.

McDonald, A. and Kay, D. (1988) *Water Resources, Issues and Strategies*, Wiley, New York.

McFeters, G.A. (1990) Enumeration, occurrence, and significance of injured indicator bacteria in drinking water, in *Drinking Water Microbiology*, (ed. G.A. McFeters), Springer, New York, pp. 478–92.

McFeters, G.A. and Camper, A.K. (1983) Enumeration of indicator bacteria exposed to chlorine. *Advances in Applied Microbiology*, 29, 177–93.

McFeters, G.A. and Singh, A. (1992) Detection methods for waterborne pathogens, in *Environmental Microbiology*, (ed. R. Mitchell), Wiley, New York, pp. 125–57.

McFeters, G.A., Bissonette, G.K. and Jezeski, J.J. (1974) Comparative survival of indicator bacteria and enteric pathogens in well water. *Applied Microbiology*, 27, 823–9.

McFeters, G.A., Cameron, S.C. and Le Chevallier, M.W. (1982) Influence of diluents, media and membrane filters on detection of injured waterborne coliform bacteria. *Applied and Environmental Microbiology*, 43, 97–103.

McFeters, G.A., Le Chevallier, M.W., Singh, A. and Kippen, J.S. (1986) Health significance and occurrence of injured bacteria in drinking water. *Water Science and Technology*, 10, 227–31

McFeters, G.A., Pyle, B.H., Gillis, S.J. *et al.* (1993) Chlorine injury and the comparative performance of Colisure, Colilert and Coliquick for the enumeration of coliform bacteria and *E. coli* in drinking water. *Water Science and Technology*, 27, 261–5.

McGowan, K., Wickersham, F. and Strockbine, N. (1989) *E. coli* 0157:H7 from water. *Lancet*, 1, 967–8.

Melnick, J.L. and Gerba, C.P. (1980) Viruses in water and soil. *Public Health Reviews*, 9, 185–213.

Mentzing, L.O. (1981) Waterborne outbreaks of *Campylobacter* enteritis in central Sweden. *Lancet*, 2, 352–4.

Metcalf, T.G. (1978) Indicators for viruses in natural waters, in *Water Pollution Microbiology*, vol. 2., (ed. R. Mitchell), Wiley-Interscience, New York, pp. 301–25.

Miyabara, Y., Sakata, Y. Suzuki, J. and Suzuki, S. (1994) Estimation of faecal pollution based on the amounts of urobilins in urban rivers. *Environmental Pollution*, 84, 117–22.

Moore, B. (1971) Health hazards of pollution, in *Microbial Aspects of Pollution*, (eds G. Sykes and F.A. Skinner), The Society for Applied Bacteriology Symposium, Series no. 1, Academic Press, London, pp. 11–33.

Moore, J.C. (1989) *Balancing the Needs of Water Use*, Springer-Verlag, New York.

Morinigo, M.A., Cornax, R., Munoz, M.A. *et al.* (1990) Relationships between *Salmonella* spp. and indicator microorganisms in polluted natural waters. *Water Science and Technology,* 24, 117–20.

Morinigo, M.A., Wheeler, D., Berry, C. *et al.* (1992) Evaluation of different bacteriophage groups as faecal indicators in contaminated natural waters in south England. *Water Research,* 26, 267–71.

Morinigo, M.A., Martinez-Manzanares, M.A., Munoz, M.A. *et al.* (1993) Reliability of several micro-organisms to indicate the presence of *Salmonella* in natural waters. *Water Science and Technology,* 27, 471–4.

Morrison, A. (1983) In third world villages, a simple handpump saves lives. *American Society of Civil Engineering,* 52, 68–72.

Mossel, D.A. (1982) Marker (index and indicator) organisms in food and drinking water. Semantics, ecology, taxonomy and enumeration. *Antonie van Leeuwenhoek,* 48, 608–11.

Moynihan, M. (1992) *Emerging Issues for the Microbiology of Drinking Water,* MSc thesis, University of Dublin.

Mukhopadhyay, A.K., Saha, P.K., Garg, S. *et al.* (1995) Distribution and virulence of *Vibrio cholerae* belonging to serotypes other than 01 and 0139: A nationwide survey. *Epidemiology and Infection,* 114, 65–70.

Munoa, F.J. and Panes, R. (1988) Selective medium for isolation and enumeration of *Bifidobacterium* spp. *Applied and Environmental Microbiology,* 54, 1715–18.

Musial, C.E., Arrowood, M.J., Sterling, C.R. and Gerba, C.P. (1987) Detection of *Cryptosporidium* in water using polypropylene cartridge filters. *Applied and Environmental Microbiology,* 53, 687–92.

Nair, S., Poh, C.L., Lim, Y.S. *et al.* (1994) Genome finger printing of *Salmonella typhi* by pulsed field gel electrophoresis for sub-typing common phage types. *Epidemiology and Infection,* 113, 391–402.

Nasser, A.M., Tchorch, Y. and Fattal, B. (1993) Comparative survival of *E. coli,* F+ bacteriophages, HAV and Poliovirus 1 in wastewater and groundwater. *Water Science and Technology,* 27, 401–7.

Neill, M.A. (1994) *E. coli* 0157:H7 time capsule: what did we know and when did we know it? *Dairy, Food and Environmental Sanitation,* 14, 374–7.

Ohasi, M. (1988) Typhoid fever, in *Laboratory Diagnosis of Infectious Disease Principles and Practices. Bacterial Mycotic and Parasitic Diseases,* vol. 1, (eds A. Barlows *et al.*), Springer–Verlag, Berlin, pp. 525–32.

Oliveri, V.P. (1982) Bacterial indicators of pollution, in *Bacterial Indicators of Pollution,* (ed. W.O. Pipes), CRC Press, Boca Raton, FL, pp. 21–42.

Olson, B.H. and Nagy, L.A. (1984) Microbiology of potable water. *Advances in Applied Microbiology,* 30, 73–132.

Olson, B.H., Clark, D.L., Milner, B.B. *et al.* (1991) Total coliform detection in drinking water – comparison of membrane filtration with Colilert and Coliquick. *Applied and Environmental Microbiology,* 57, 1535–9.

Ongerth, J.E. and Stibbs, H.H. (1987) Identification of *Cryptosporidium* oocysts in river water. *Applied and Environmental Microbiology,* 53, 672–76.

Packham, A.F. (1990) *Cryptosporidium* and water supplies – the Badenoch Report. *Journal of the Institution of Water and Environmental Management,* 4, 578–80.

Paquin, J.L., Block, J.C., Haudidier, K. *et al.* (1992) Effect of chlorine on the bacterial colonization of a model distribution system. *Revue des Sciences de L'eau,* 5, 399–414.

Parker, J.F., Greaves, G.F. and Smith, H.V. (1993) The effect of ozone on the viability of *Cryptosporidium parvum* oocysts and a comparison of experimental methods. *Water Science and Technology*, 27, 93–6.

Patterson, W.J., Seal, D.V., Curran, E. *et al.* (1994) Fatal nosocomial Legionnaires' disease: relevance of contamination of hospital water supplies by temperature-dependent buoyancy-driven flow from spur pipes. *Epidemiology and Infection*, 112, 513–25.

Pavlova, M.F., Brezenski, F.T. and Litsky, W. (1972) Evaluation of various media for isolation, enumeration and identification of faecal streptococci from natural sources. *Health Laboratory Sciences*, 9, 289–98.

Payment, P. (1991) Elimination of coliphage, *Clostridia perfringens* and human enteric viruses during drinking water treatment: results of large volume samplings. *Water Science and Technology*, 24, 213–5.

Payment, P., Franco, E. and Siemiatycki, J. (1993) Absence of a relationship between health effects due to tap water consumption and drinking water quality parameters. *Water Science and Technology*, 27, 137–45.

Payment, P., Eduardo, F. Richardson, L. and Siemiatycki, J. (1991) Gastrointestinal health effects associated with the consumption of drinking water produced by point of use domestic reverse osmosis filtration units. *Applied and Environmental Microbiology*, 57, 945–8.

Peeters, J.E., Opdenbosch, E. van and Glorieux, B. (1989) Effect of disinfection of drinking water with ozone or chlorine dioxide on survival of *Cryptosporidium parvum* oocysts. *Applied and Environmental Microbiology*, 55, 1519–22.

Peterson, D.A., Hurley, T.R., Hoff, J.C. and Hoff, L.G. (1983) Effect of chlorine treatment on infectivity of Hepatitis A virus. *Applied and Environmental Microbiology*, 45, 223–7.

Petzel, J.P. and Hartman, P.A. (1985) Monensin-based medium for determination of total gram negative bacteria and *E. coli*. *Applied and Environmental Microbiology*, 49, 925–33.

Philipp, R. (1991) Risk assessment and microbiological hazards associated with recreational water sports. *Reviews in Medical Microbiology*, 2, 208–14.

PHLS/SCA (Public Health Laboratory Service, Standing Committee on the Bacteriological Examination of Water Supplies) (1972) Comparison of membrane filtration and multiple tube methods for the enumeration of coliform organisms in water. *Journal of Hygiene*, 70, 691–705.

PHLS/SCA (Public Health Laboratory Service, Standing Committee of Analysts) (1980a) A comparison between minerals-modified glutamate medium and lauryl tryptose lactose broth for the enumeration of *E. coli* and coliform organisms in water by the multiple tube method. *Journal of Hygiene*, 85, 35–48.

PHLS/SCA (Joint Committee of Public Health Laboratory Service and Standing Committee of Analysts) (1980b) Membrane filtration media for the enumeration of coliform organisms and *E.coli* in water: comparison of Tergitol 7 and Lauryl Sulphate with Teepol 610. *Journal of Hygiene*, 85, 181–91.

Pipes, W.O. (1982a) Indicators and water quality, in *Bacterial Indicators of Pollution*, (ed. W.O. Pipes), CRC Press, Boca Raton, FL, pp. 83–96.

Pipes, W.O. (1982b) Introduction, in *Bacterial Indicators of Pollution*, (ed. W.O. Pipes), CRC Press, Boca Raton, FL, pp. 1–20.

Pipes, W.O., Minnigh, H.A., Moyer, B. and Troy, M.A. (1986) Comparison of Clark's P–A test and the membrane filter method for coliform detection in potable water samples. *Applied and Environmental Microbiology*, 52, 439–43.

Pochin, A. (1975) The acceptance of risk. *British Medical Journal*, 31, 184–8.

Pollitzer, R. (1959) *Cholera*, WHO Monograph Series 43, World Health Organization, Geneva.

Pontius, F.W. (1993) Federal Drinking Water Regulations update. *Journal of the American Water Works Association*, 85, 43.

Poucher, A.M., Devriese, L.A., Hernandez, J.F. and Delattre, J.M. (1991) Enumeration by a miniaturised method of *E. coli, Streptococci bovis* and *Enterococci* as indicators of the origin of faecal pollution of waters. *Journal of Applied Bacteriology*, 70, 525–30.

Pöyry, T., Stenvik, M. and Hovi, T. (1988) Viruses in sewage waters during and after a polio myelitis outbreak and subsequent nationwide oral poliovirus vaccination campaign in Finland. *Applied and Environmental Microbiology*, 54, 371–4.

Prescott, L.M., Harley, J.P. and Klein, D.A. (1993) *Microbiology*, WCB Publishers, Iowa.

Presswood, W.G. and Brown, L.R. (1973) Comparison of Gelman and Millipore membrane filters for enumerating faecal coliform bacteria. *Applied Microbiology*, 26, 332–6.

Ramteke, P.W., Bhattacharjee, J.W., Pathak, S.P. and Kaira, N. (1992) Evaluation of coliforms as indicators of water quality in India. *Journal of Applied Bacteriology*, 72, 352–6.

Ratto, A., Dutka, B.J., Vega, C. *et al.* (1989) Potable water safety assessed by coliphage and bacterial tests. *Water Research*, 23, 253–5.

Reasoner, D.J. (1988) Drinking water microbiology, research in the United States: an overview of the past decade. *Water Science and Technology*, 20, 101–7.

Reasoner, D.J. (1992) *Pathogens in Drinking Water – Are There Any New Ones?* US EPA, Washington, DC.

Reasoner, D.J. and Geldreich, E.E. (1985) A new medium for the enumeration and subculture of bacteria from potable water. *Applied and Environmental Microbiology*, 49, 1–7.

Reasoner, D.J., Blannon, J.C. and Geldreich, E.E. (1979) Rapid 7 hour faecal coliform test. *Applied and Environmental Microbiology*, 38, 229–36.

Reilly, W.J. (1995) *E. coli* 0157 – The Scottish situation, in *Coliforms and* E. coli: *Problem or Solution*, abstract of papers and posters of conference held at University of Leeds, September, 1995.

Resnick, I.G. and Levin, M.A. (1981) Assessment of *Bifidobacteria* as indicators of human faecal pollution. *Applied and Environmental Microbiology*, 42, 433–8.

Rice, E.W., Geldreich, E.E. and Read, E.J. (1989) The P–A coliform test for monitoring drinking water quality. *Public Health Report*, 104, 54–8.

Richards, J.C.S., Jason, A.C., Hobbs, G. *et al.* (1978) Electronic measurement of bacterial growth. *Journal of Physics, E: Scientific Instruments*, 11, 560–8.

Richardson, K.J., Stewart, M.H. and Wolfe, R.L. (1991) Application of gene probe technology for the water industry. *Journal of the American Water Works Association*, 83, 71–81.

Ridgeway, H.F. and Olson, B.H. (1981) Microscopic evidence for bacterial colonisation of a drinking water distribution system. *Applied and Environmental Microbiology*, 42, 274–87.

Ridgeway, H.F., Justice, C.A., Wittaker, C. *et al.* (1984) Biofilm fouling of RO membranes – its nature and effect on treatment of water for reuse. *Jounal of the American Water Works Association*, 77, 94–102.

Riley, W.J., Forbes, G.I., Paterson, G.M. and Sharp, J.C.M. (1981) Human and animal Salmonellosis in Scotland associated with environmental contamination 1973–1979. *Veterinary Record*, 37, 553–5.

Rivera, S.C., Hazen T.C. and Toranzos, G.A. (1988) Phenotypic and genotypic comparison of *E. coli* from pristine tropical waters. *Applied and Environmental Microbiology*, 54, 979–83.

Robinton, E.D. and Mood, E.W. (1966) A quantitative and qualitative appraisal of microbial pollution of water by swimmers: a preliminary report. *Journal of Hygiene*, 69, 489–99.

Rose, J.B. (1988) Occurrence and significance of *Cryptosporidium* in water. *Journal of the American Water Works Association*, 80, 53–8.

Rose, J.B. (1990) Emerging issues for the microbiology of drinking water. *Engineering and Management*, 90(7), 23–9.

Rose, J.B. and Gerba, C.P. (1991) Use of risk assessment for development of microbial standards. *Water Science and Technology*, 24, 29–34.

Rose, J.B., Darbin, H. and Gerba, C.P. (1988) Correlations of the protozoa *Cryptosporidium* and *Giardia* with water quality variables in a watershed. *Water Science and Technology*, 20, 271–6.

Rosenfeld, W.D. and Zobell, C.E. (1947) Antibiotic production by marine microorganisms. *Journal of Bacteriology*, 54, 393–8.

Roszak, D.B. and Colwell, R.R. (1987) Survival strategies of bacteria in the natural environment. *Microbiological Reviews*, 51, 365–79.

Rowbotham, T.J. and Cross, T. (1977) *Rhodococcus coprophilus* spp. nov.: An aerobic Nocardioform Actinomycete belonging to the 'Rhodochrous' complex. *Journal of General Microbiology*, 100, 123–38.

Rutkowski, A.A. and Sjøgren, R.E. (1987) Streptococcal population profiles as indicators of water quality. *Water, Air and Soil Pollution*, 34, 273–84.

Safe Drinking Water Committee (1977) *Drinking Water and Health*, National Academy of Sciences, Washington DC.

Saiki, R.K., Gelfad, D.H., Stoffel, S. *et al.* (1988) Primer directed enzymatic amplification of DNA with a thermostable DNA polymerase. *Science*, 239, 487–94.

Samonis, G., Elting, L., Skoulika, E. *et al.* (1994) An outbreak of diarrhoeal disease attributed to *Shigella sonnei*. *Epidemiology and Infection*, 112, 235–45.

Santiago-Mercado, J. and Hazen, T.C. (1987) Comparison of four membrane filter methods for faecal coliform enumeration in tropical waters. *Applied and Environmental Microbiology*, 53, 2922–8.

Sato, M.T.Z., Sanchez, P.S., Alves, M.N. *et al.* (1995) Evaluation of culture media for *Candida albicans* and *Staphylococcus aureus* recovery in swimming pools. *Water Research*, 29, 2412–16.

Sauch, J.F. (1986) *Giardia Detection in Water Supplies*, Health Effects Research Laboratory, US EPA, EPA/600/D – 86/234.

Sayler, G.S. and Layton, A.C. (1990) PCR applications in environmental microbiology. *Annual Review of Microbiology*, 45, 137–61.

Schbert, R.H. (1991) Aeromonads and their significance as potential pathogens in water. *Journal of Applied Bacteriology* (Symposium Supplement), 70, 131–5.

Seidler, R.J., Evans, T.M., Kaufman, J.R. *et al.* (1981) Limitations of standard coliform enumeration techniques. *Journal of the American Water Works Association*, 73, 538–42.

Sellwood, J. and Dadswell, J. (1991) Human viruses and water, in *Current Topics in Clinical Virology*, (ed. P. Morgan-Caprier), Laverham Press, Salisbury, pp. 29–45.

Shaffer, P.T.B., Metcalf, T.G. and Sproul, O.J. (1980) Chlorine resistance of coliform isolates recovered from drinking water. *Applied and Environmental Microbiology*, 40, 1115–21.

Short, C.S. (1988) The Bramham incident, 1980 – an outbreak of waterborne infection. *Journal of the Institution of Water and Environmental Management*, 2, 383–90.

Shuval, H.I., Cohen, J. and Kolodney, R. (1973) Regrowth of coliforms and faecal coliforms in chlorinated wastewater effluent. *Water Research*, 7, 537–46.

Simkova, A. and Cervenka, J. (1981) Coliphage as ecological indicators of enteroviruses in various water systems. *Bulletin of the World Health Organisation*, 4, 611–8.

Singh, A. and McFeters, G.A. (1992) Detection methods for waterborne pathogens, in *Environmental Microbiology* (ed. R. Mitchell), Wiley, New York.

Sinton, L.W., Donnison, A.M. and Hastie, C.M. (1993a) Faecal streptococci as faecal pollution indicators: a review. Part I: Taxonomy and enumeration. *New Zealand Journal of Marine and Freshwater Research*, 27, 101–15.

Sinton, L.W., Donnison, A.M. and Hastie, C.M. (1993b) Faecal streptococci as faecal pollution indicators: a review. Part II: Sanitary significance, survival and use. *New Zealand Journal of Marine and Freshwater Research*, 27, 117–37.

Skirrow, M.B. and Blaser, M.J. (1992) Clinical and epidemiological considerations, in *Campylobacter jejuni, Current Status and Future Trends*, (eds I. Nachamkin, M.J. Blaser and L.S. Tomkins), American Society for Microbiology, Washington, DC, pp. 3–8.

Sladek, K.J., Suslavich, R.V., Isom, B. and Dawson, F.W. (1975) Optimum membrane structures for growth of coliform and faecal coliform organisms. *Applied Microbiology*, 30, 685.

Smith, D.B., Hess, A.F. and Hubbs, S.A. (1989) Survey of distribution system coliforms in the United States, in *Proceedings of 1989 AWWA Water Quality Technology Conference, San Diego*, American Water Works Association, Washington, DC, pp. 1103–15.

Smith, H.V. (1992) Cryptosporidium and water: a review. *Journal of the Institute of Water Engineers and Scientists*, 6, 443–51.

Smith, H.V. and Rose, J.B. (1990) Waterborne Cryptosporidiosis. *Parasitology Today*, 6, 8–12.

Snyder, J.D. and Merson, M.H. (1982) The magnitude of the global problem of acute diarrhoeal disease: a review of active surveillance data. *Bulletin of the World Health Organisation*, 60, 603–13.

Sobsey M.D. (1975) Enteric viruses and water supplies. *Journal of the American Water Works Association*, 67, 414–18.

Sobsey, M.D. (1989) Inactivation of health related microorganisms in water by disinfection processes. *Water Science and Technology*, 21, 179–95.

Sobsey, M.D., Dufour, A.P., Gerba, C.P. *et al.* (1993) Using a conceptual framework for assessing risks to human health from microbes in drinking water. *Journal of the American Water Works Association*, 85, 44–8.

Sorvillo, F., Lieb, L.E., Nahlen, B. *et al.* (1994) Municipal drinking water and cryptosporidiosis among persons with AIDS in Los Angeles County. *Epidemiology and Infection*, 113, 313–20.

Standing Committee of Analysts (1990) *Identification of Giardia Cysts, Cryptosporidium Oocysts and Free Living Pathogenic Amoebae in Water etc.*, HMSO, London.

States, S.J., Wadowsky, R.M., Kuchta, J.M. *et al.* (1990) *Legionella* in drinking water, in *Drinking Water Microbiology*, (ed. G.A. McFeters), Springer–Verlag, New York, pp. 340–67.

Steffan, R.J. and Atlas, R.M. (1991) Polymerase chain reaction: applications in environmental microbiology. *Annual Review of Microbiology*, 45, 137–61.

Stetler, R.E. (1984) Coliphage as indicators of enteroviruses. *Applied and Environmental Microbiology*, 48, 668–70.

Sticht-Groh, V. (1982) *Campylobacter* in healthy slaughter of pigs: a possible source of infection of man. *Veterinary Record*, 110, 104–6.

Stout, J., Yu, V.L. and Best, M.G. (1985) Ecology of *Legionella pneumophilia* within water distribution systems. *Applied and Environmental Microbiology*, 49, 584.

Stuart, D.A., McFeters, G.A. and Schillinger, J.E. (1977) Membrane filter technique for the quantification of stressed faecal bacteria in the aquatic environment. *Applied and Environmental Microbiology*, 34, 42–6.

Swedlow, D.L. and Ries, A.A. (1993) *Vibrio cholerae* non-01: the eighth pandemic? *Lancet*, 342, 382–3.

Tartera, C. and Jofre, J. (1985) *Bacteroides* bacteriophage as potential indicators of human viruses in the environment. *Abstracts of the Annual Meeting of the American Society for Microbiologists*, 270.

Tartera, C. and Jofre, J. (1987) Bacteriophage active against *Bacteroides fragilis* bacteriophage as indicators of the viriological quality of water. *Water Science and Technology*, 18, 1623–37.

Tartera, C., Lucena, F. and Jofre, J. (1989) Human origin of *Bacteroides fragilis* bacteriophage present in the environment. *Applied and Environmental Microbiology*, 55, 2696–701.

Tauxe, R.V. (1992) Epidemiology of *Campylobacter jejuni* infections in the US and other industrialised nations, in *Campylobacter jejuni, Current Status and Future Trends*, (eds I. Nachamkin, M.J. Blaser and L.S. Tomkins), American Society for Microbiology, Washington, DC, pp. 9–19.

Taylor, D.N., McDermott, K.T., Little, J.R. *et al.* (1983) *Campylobacter* enteritis from untreated water in the Rocky Mountains. *American Internal Medicine*, 99, 38–40.

Tebutt, T.H.Y. (1992) *Principles of Water Quality Control*, Pergamon Press, Oxford.

Thomas, H.A. Jr., Woodward, R.L. and Kable, P.W. (1956) Use of molecular filter membrane for water quality control. *Journal of the American Water Works Association*, 48, 1391–402.

Tillett, H.E. and Coleman, R. (1985) Estimated numbers of bacteria in samples from non-homogeneous bodies of water: How should MPN and membrane filtration results be reported. *Journal of Applied Bacteriology*, 59, 381–8.

Tillett, H.E. and Farrington, C.P. (1991) Inaccuracy of counts of organisms in water or other samples: effects of predilution. *Letters in Applied Microbiology*, 13, 168–70.

Tobin, L.S. and Dutka, B.J. (1977) Comparison of the surface structure, metal binding and faecal coliform recoveries of nine membrane filters. *Applied and Environmental Microbiology*, 34, 69–79.

Tobin, L.S., Lomax, J. and Kushner, D.J. (1980) Comparison of nine brands of membrane filtration and the most probable number methods for total coliform enumeration in sewage contaminated drinking water. *Applied and Environmental Microbiology*, 40, 186–91.

Toranzos, G.A. (1991) Current and possible alternate indicators of faecal contamination in tropical waters: a short review. *Environmental Toxicology and Water Quality*, 6, 121–30.

Townsend, S.A. (1992) The relationships between *salmonella* and faecal indicator concentrations in 2 pools in the Australian wet/dry tropics. *Journal of Applied Bacteriology*, 73, 182–8.

Tsai, Y.L., Plamer, C.J. and Sangermano, L.R. (1993) Detection of *E. coli* in sewage and sludge by polymerase chain reactions. *Applied and Environmental Microbiology*, 59, 353–7.

Tyler, J. (1985) Occurrence in water of viruses of public health significance. *Journal of Applied Bacteriology*, (Symposium Supplement), 37s–46s.

US EPA (1976) *National Interim Primary Drinking Water Regulations*, EPA – 570/9–76–003, Washington, DC.

US EPA (1990a) *Risk Assessment, Management and Communication of Drinking Water Regulation*, EPA/625/4–89/024, US EPA, Cincinnati, OH.

US EPA (1990b) *Drinking Water Regulations under the Safe Drinking Water Act*, fact sheet, May, US EPA.

US EPA (1992) *National Primary Drinking Water Regulations – Analytical Techniques Coliform Bacteria, Final Rule*. Federal Register, 57, 24744–7.

US EPA (1993) *Drinking Water Regulations and Health Advisories*, Health and Ecological Criteria Division, US EPA, Washington, DC.

Van Poucke, S.O. and Nellis, H.J. (1995) Detection of coliforms by chemiluminometric measurement of ß-galactosidase activity, in *Coliforms and E. coli: Problem or Solution*, abstract of papers and posters of conference held at University of Leeds, September, 1995.

Vascoucelos, G.J. and Swartz, R.G. (1976) Survival of bacteria in seawater using a diffusion chamber apparatus *in situ. Applied and Environmental Microbiology*, 31, 913–20.

Versteagh, J.F., Haavelaar, A.H., Hoekstra, A.C. and Visser, A. (1989) Complexing of copper in drinking water samples to enhance recovery of *Aeromonas* and other bacteria. *Journal of Applied bacteriology*, 67, 561–6.

Vogt, R.L. *et al.* (1982) *Campylobacter* enteritis associated with contaminated water. *Annals of Internal Medicine*, 96, 292–6.

WAA (1985) *Guide to the Microbiological Implications of Emergencies in the Water Services*, Water Authorities Association, London.

WAA (1988) *Operational Guidelines for the Protection of Drinking Water Supplies: Safeguards in the Operation and Management of Public Water Supplies in England and Wales*, Water Authorities Association, London.

Waite, T.D. (1984) *Principles of water quality*, Academic Press, London.

Waite, W.M. (1985) A critical appraisal of the coliform test. *Journal of the Institute of Water Engineers and Scientists*, 39, 341–57.

Walker, R.W., Wun, C. and Litsky, W. (1982) Coprostanol as an indicator of fecal pollution. *CRC Critical Reviews in Environmental Control*, 10, 91–112.

Watanabe, T., Shimohashi, H., Kawai, Y. and Mutal, M. (1981) Studies on *Streptococcus* 1: distribution of faecal streptococci in man. *Microbiology and Immunology*, 25, 257–69.

Watkins, J. and Cameron, S.A. (1991) Recently recognized concerns in drinking water microbiology. *Journal of the Institution of Water and Environmental Management*, 5, 624–30.

Wende, J. van der, Characklies, W.G. and Smith, D.B. (1989) Biofilms and bacterial drinking water quality. *Water Research*, 23, 1313.

West, P.A. (1991) Human pathogenic viruses and parasites: Emerging pathogens. *Journal of Applied Bacteriology* (Symposium Supplement), 70, 1075–145.

Wheater, D.W.F., Mara, D.D. and Oragui, J. (1979) Indicator systems to distinguish sewage from stormwater run off and human from animal faecal material, in *Biological Indicators of Water Quality*, (eds A. James and L. Evison), Wiley, Chichester, pp. 21/1–27.

White, S.F. and Mays, G.D. (1989) *Water Supply Bylaws Guide*, 2nd edn, Water Research Centre and Ellis Horwood, Chichester.

WHO (1984) *Guidelines for Drinking Water Quality*, vol. 2: *Health Criteria and Other Supporting Information*. World Health Organization, Geneva.

WHO (1993) *Guidelines for Drinking Water Quality*, 2nd edn, vol. 1: *Recommendations*, World Health Organization, Geneva.

Wierenga, J.T. (1985) Recovery of coliforms in the presence of free chlorine residual. *Journal of the American Water Works Association*, 78, 83–8.

Wilkins, M. (1993) The fight against Cryptosporidium. *Water Bulletin*, 544, 12.

Windle-Taylor, E. and Burman, N.P. (1964) The application of membrane filtration techniques to the bacteriological examination of water. *Journal of Applied Bacteriology*, 27, 294–303.

Wolf, H.W. (1972) The coliform count as a measure of water quality, in *Water Pollution Microbiology*, vol. 1, (ed. R. Mitchell), Wiley-Interscience, New York, pp. 333–45.

Wong, N.A., Linton, C.J., Jalal, H. and Millar, M.R. (1994) Randomly amplified polymorphic DNA typing: a useful tool for rapid epidemiological typing of *Klebsiella pneumoniae*. *Epidemiology and Infection*, 113, 445–54.

Wong, S.S.Y., Yuen, K.Y., Yam, W.C. et al. (1994) Changing epidemiology of human salmonellosis in Hong Kong, 1982–93. *Epidemiology and Infection*, 113, 425–34.

Woodward, R.L. (1957) How probable is the Most Probable Number? *Journal of the American Water Works Association*, 49, 1060–8.

WRc (1991) *Recovery of Cryptosporidium From Water*, FR 0189, Foundation for Water Research, Marlow.

WRc (1994) *Removal of Cryptosporidium Oocysts by Water Treatment Processes*, FR 0457, Foundation for Water Research, Marlow.

Wright, E.P. (1982) The occurrence of *Campylobacter jejuni* in dog faeces from a public park. *Journal of Hygiene*, Cambridge, 89, 191–4.

Wright, R.C. (1982) A comparison of the levels of faecal indicator bacteria in water and human faeces in a rural area of a tropical developing country (Sierra Leone) *Journal of Hygiene*, Cambridge, 89, 69–78.

Writer, J.H., Leenheer, J.A., Barber, L.B. et al. (1995) Sewage contamination in the Upper Mississippi River as measured by fecal sterol, coprostanol. *Water Research*, 29, 1427–36.

Yamamato, T., Morotomi, M. and Tanaka, R. (1992) Species specific oligonucleotide probes for five *Bifidobacteria* spp. detected in human intestinal microflora. *Applied and Environmental Microbiology*, 58, 4076–9.

Yoshpe-Purer, Y. (1989) Evaluation of media for monitoring fecal streptococci in seawater. *Applied and Environmental Microbiology*, 55, 2041–5.

Yoshpe-Purer, Y. and Golderman, S. (1987) Occurrence of *Staphylococcus aureus* and *Pseudomonas aeriginosa* in Israeli coastal waters. *Applied and Environmental Microbiology*, 53, 1138–41.

Zaske, S.K., Dockins, W.S. and McFeters, G.A. (1980) Cell envelope damage in *E. coli* caused by short term stress in water. *Applied and Environmental Microbiology*, 40, 386–90.

Index